My dog can't read ...

My dog can't read ...

More tales from an accidental farmer

Diana Ashworth

LOGASTON PRESS

First published in 2025 by Logaston Press,
The Holme, Church Road, Eardisley HR3 6NJ.
www.logastonpress.co.uk
An imprint of Fircone Books Ltd.

ISBN 978-1-910839-82-9

Designed and typeset by Richard Wheeler in 11 on 14 Caslon.
Cover designed by Richard Wheeler.
Printed and bound in Poland www.lfbookservices.co.uk

Logaston Press is committed to a sustainable future for our business, our readers and our planet. The book in your hands is made of FSC® certified and other controlled material.

British Library Catalogue in Publishing Data.
A CIP catalogue record for this book is available from the British Library.

CONTENTS

With thanks to Gay Roberts, Norma Allen and Chris Barrett for recognising the potential of these stories and making me write them. To Peter Carr, Su and Richard Wheeler for their invaluable editorial advice, and to my neighbours, friends and family for their inspiration.

Survival?

I DIDN'T SEE THE ACCIDENT, but I do notice the blood on the road as I swerve to miss her as she staggers blindly in my way. I stop, and the cars behind are already pulling out to overtake me. I switch on the hazard lights and scramble over to jump out from the passenger door and run back.

No one else stops – they don't even see her – now she lies helpless in the deep gutter. I must be quick or there will be another accident.

I can smell her blood – it is on my hands as I try to hold her, steady her. She struggles and kicks. There are two huge gashes on her head – I can see the bone. Liquid is bubbling from one of her eyes and blood is coming from her nostrils. Strangely the cars whizz past, their drivers oblivious to the drama.

I have nothing with which to do anything. I run back, open the car boot and find only some carrier bags and the dog lead (no first aid kit of course – anyway there is no time for that). I tear the bags flat and wrap her in them, swaddling this little creature like a baby, and truss her up with the dog lead so that she cannot injure herself anymore. I talk quietly to her as I do this and she calms. I lay her in the dark boot of the car, wedging her in so she won't roll about. Then I close the lid with an ominous clunk.

Now, I could drive into town, to the vet – yes, she is a duck, a little mallard, hit by a car. Well, she probably flew into the moving vehicle, but I do not take her to the vet. I've seen the expression on their faces when you present a country practitioner with a wild thing (and I've paid the price). No, I take her home – for life is life and death is death, regardless of the species.

My husband groans, and once again our newly-tiled wet-room comes into its own. It spends far more hours nursing the sick than it does washing our family and friends.

Trying to walk, she repeatedly topples over to the right, but in the half-light of the darkened shower room she settles and sits quietly all that day and all the next. Nothing eats her. She moves around a little when no one is looking, but will not eat the small slugs which I have collected for her and which climb their slimy way circuitously to the ceiling. Nor does she try the bread in water which she spills, nor the caterpillar that pupates on the tap.

On the third day, she is thin, dehydrated and matted but walks more steadily and looks up at me with her one good eye as if she sees me. We have to go away to a funeral the next day, so first thing in the morning I carry her to our pond and put her down gently by its edge. All the way there she is looking from side to side as if getting her bearings. Next thing, she topples forward and *Plop!* – she's in the water. She lowers her head so that the pond water flows into her bill, and she takes a long, cool drink and paddles off purposefully. On the far side she climbs out onto the bank under the muddy cliff where the water from the spring runs down in a curtain. She settles there, washed by the tiny waterfall.

Next evening when we return she is still there, beside but not under the waterfall. She watches me throw bread onto the water then stands up straight and flaps her wings two or three times to test them, shakes herself and settles down again.

Next morning the bread is gone and so is she, flown away or carried off by a fox. But wait ... there she is, sitting near the path, ready for breakfast.

We have a friend who is a farmer: when asked if he has to get up a lot in the night for his animals he says, 'No, God does the night shift'.

~

The Trannon Valley is just about the most peaceful place you could find. The most violent living things, the trees, with all their pent-up energy, have been known to lash out under provocation from a chainsaw, and knock a man off a step ladder or into the river. A herd of cattle once wandered onto the road and was alarmed by a stranger's car so that they stampeded up and down the steeply-banked lane. Generally though, ask any young resident, it is excruciatingly peaceful.

Today, as the sun comes out, a lady in the village decides to have a clear out. Her late husband had been a hoarder by all accounts, and a practical joker, as was his father. She is sorting through the memories – the wrong-size golf clubs and walking sticks, a pith helmet, belts and brasses and carved knick-knacks, a box of strange-looking bulbs that don't fit any known socket, and keys with no locks, of cars long-sold and of cases left in foreign parts. Here's a brand-new cricket ball, and what's this? It looks like a hand-grenade ... It *is* a hand-grenade!

Now the lady is very sensible – unflappable. She doesn't throw open the bedroom window and lob the grenade into the field behind the house (where a grazing cow of a curious and determined disposition could pull out the pin a few moments later). Neither does she do what I would do, which is put it in the bread bin or the oven (where we put all precious things that cats and dogs are not to chew) and flee. No, she stays calm and looks around in case it has a mate (she's lived in Africa). And, sure enough, there it is in the bottom of the box. It's a mystery how she knows there might be two as she had never seen them before. Then, with all the sangfroid of a lady who has slept with a pair of grenades under her bed, she calmly telephones the police.

This is when things get exciting and how the cordons and roadblocks and exclusion zone around our sleepy village arrive (although they miss Wenona, next door, who is having a nap and doesn't hear them knock). You can't blame them for getting carried away – it is enormous fun for the entire constabulary, a team-building day out, and one thing we do well in Wales is *cups of tea and cake*. Also, there aren't many officers in Mid Wales and they don't get out much.

A local landowner is consulted, and a site is chosen for the controlled detonation. This enterprising farmer dismisses the opportunity of sorting out the badgers once and for all (perhaps too near his nephew's house). But how about a pond? – How big will the crater be? – What's that in gallons? Perhaps we could do one a little higher and the other down there and then we could have a waterfall; maybe even a turbine. We could claim subsidies for habitat creation *and* green energy ...

Meanwhile, the Bomb Disposal team are rumbling up the byways from Cheltenham – or was it Colchester? (it is a long way away).

To everyone's great disappointment (except perhaps the lady and Wenona when she wakes) the experts, having made their examination, proclaim the grenades are dud – drilled and drained and filled with cooking oil by a spoilsport or a practical joker sometime between 1915 and now.

And somewhere high above our sleepy village someone is chuckling.

Echoes from the past

'In the olden days ...', as my granny used to say, you got everything you needed locally.

We've recently heeded the Welsh Assembly Government's injunction to farmers to renovate *traditional farm buildings*.

Traditionally, a hill farmer, with 12 children and as many useful acres, built with what was readily available: stone from the river bed, oak and hazel from the woods, and rushes or bracken to thatch, all lashed together with honeysuckle twine, or tiled with flat pieces of the local stone (almost slate). In the last 100 years the philosophy hasn't changed, but available materials have.

Welsh nouns have a gender and *she*, our barn, is made of scavenged materials – retired telegraph poles, tree trunks, branches, corrugated iron and galvanized sheets. There are old doors (wooden, fridge and car), a draining board and fencing panels. There are bits of garden shed – or maybe it's part of the old privy. There are tarpaulins, a First World War great coat and carpet (not on the floor, of course, but stapled onto the walls with sacks made of plastic, hessian and paper) and all held together with wire and baler twine of plastic or hemp.

It is a basic timber frame construction, clad inside and out with the aforementioned materials, and the wall cavity filled with balls of twine, sheep's wool, oily rags, jam-jars of unidentified substances, fag packets and a mummified rat.

The whole thing is reinforced by 40 years of incorporated tree growth and drained by various burrows.

We discover all this as we lovingly unpick it and sort what we find – into metal scrap (£££), firewood for next winter and rubbish which we burn now. Even though lately it has been a health and safety nightmare, demolishing it is a sacrilege, an insult to those who built this monument to ingenuity, inspired in poverty, and an antidote to materialism.

At the rear is a concrete plinth, presumably for the generator that produced the electricity before 1964. It was protected by a steel door with a rusty padlock still hanging open in the hasp. Believe it or not, a brass key, hanging nearby on a nail, still operates the smooth action of the lock without so much as a squirt of penetrating oil (the brand-new padlock we put on the field gate last year had seized up by November).

Our new gate is modern but not up to the specification required to withstand a rampaging pensioner in a six-ton digger. It should be taken off and straightened and rehung by newly-welded hinges on a brand-new post without a rotten bottom – but, well, it's cold, the ground is very wet and it's getting dark.

So, in the short term we are thrown back upon a traditional method: *Baler twineology*. Actually, it's a 'fusion' technology – we use nylon twine but in the traditional way. And don't forget that nylon has memory (which is more than you can say for the farmer): the twine remembers how to undo itself, so you should lock the knots – ironically that means reef knots and not granny knots.

The farms around us are particularly tidy and I fear that they are losing the ancient skills. To understand this, you need to study our Grade-II Listed, traditional sheepfold. Note the use of growing, self-reinforcing timber: four trees growing a hurdle-width from each other. One side is formed by the fence that runs along between two of the trees. Two other sides are formed by rusty old iron gates (circa 1900) scavenged from elsewhere and tied firmly. The fourth side opens and closes on leather hinges (sections of army belt nailed into position). The gate itself was upgraded in recent decades to an aluminium bedstead. If you look very carefully at the trees you can see where, many years ago, the earlier binding of this structure has dug into the expanding trunk of the tree and become enveloped in noduled bark, each generation leaving a ring of captured hemp or wire – or, most recently, plastic twine. I and my zip-ties will eventually make our mark, a bark bulge of our own.

We try to keep these skills alive in a modern context. Another is the algae-prevention modification of our rain-water harvester. Using light-occlusive black membrane left over from the damp-proofing of the floor of the cottage, we have parcelled up the white plastic water butt and tied it up with the ubiquitous orange twine, knots locked against the weather by melting with a cigarette lighter (simple).

I am most proud of my four-minute-cratch (patent pending). A cratch is an outside hay rack. It was erected in four minutes in a hailstorm when snow was forecast. I am particularly pleased with the re-purposing of two triangular grass collection boxes from the lawn mowers to stop the sheep getting their feet stuck and injured when trying to climb in at the ends. The back is formed by the fence, the front is an angled hurdle, and the top is half the oak door from the old pig sty, all held together by – guess what? – baler twine.

~

Dolgellau livestock market is also informed by tradition. Today, full of magnificent lambs, nearly mature now, trimmed and tidy, clothed in innocence, sniffing the air and looking to their shepherd's face, which is relaxed, for clues. These are being sold as 'store-lambs', to go on to other farms, not for slaughter. The owners are easy, joking, ready to chat. Cull ewe sales are a different matter, the sheep are left and the farmers have urgent matters to attend to elsewhere. They don't hang around to see old ewes sold, those they have lambed and pulled out of bogs in the middle of winter or dug out of snow drifts.

Today, farmers perch on the pen rails in the autumn sun, and chew the cud. They look at lambs smaller than their own and bemoan the future for those farming on the edge, on the high, rugged land that you can see from the market, where the mountain pass rises by Cadair Idris.

Old men remember the winter of '47, or riding from Llanbrynmair to Aberystwyth over the moors without having to open a gate; and the sheep court at Dylife where they sorted out the stray sheep once a year.

Young men know that the openness of the country still prevents them controlling their stock as they would like. One cautious man, mindful of recent late springs, decided to lamb later (as they used to) only to find that his neighbour's tup had already serviced half his ewes!

Bold marking helps muster sheep in mountainous ground where the hardy beasts jump stone walls. Clear marks make good neighbours, hence the colourful pens of sheep.

Today, there are also some people showing their pedigree Improved Welsh Mountain Sheep. Prior to this, I had harboured hopes of one day producing a breeding ram, but I realise this is folly. These creatures have huge, curled horns. I visualise one of his ram-lambs hooking himself on a fence in the middle of the night, to wake us up; or worse, to engulf me in guilt when we find his little body hanging there next morning.

The sight of a large ram walking to heel on a lead is incongruous, like some strange dog, one that occasionally has a flash of recall and lowers its head, arches its back and kicks out its back legs, like a bison, before turning its head graciously to the camera. Breeders are flushed with pride but avoid any undue show of emotion. This is, after all, a livestock market.

It is a proper, manly place, not the sort of establishment where buyers are squeamish about testicles or undocked tails. In this more rugged terrain, there are enough challenges for young animals without adding to their stress by castrating or introducing infection. And anyway, I thought testosterone built up muscle and isn't that what it's all about? 'It's not that simple!' our young friends say.

That brings me to lunch. Just £5 for a massive bowl of casseroled lamb's liver and bacon with baked vegetables and mashed potatoes, eaten from the bowl with a knife and fork. Not because it's trendy; it's always been done that way, and fit to serve to anyone, anywhere. And pies and cakes and home-made fruit flans with cream are to die for ... But that's not a problem for the men who still run up these steep hills tending their flocks; who carry sacks of feed, three at a time, and lug fully-grown ewes about as if they are tired children, and walk rams about on leads.

~

Back home, Alan expertly backs the borrowed stock trailer into our yard, just because he can. It is dusk now, but we can see the ewes gathered in a great arc, all watching something in the dip. We go to investigate and see dim lights and movement in the road below the farm, where three ways meet. This is where there is an ancient lorry trap. It used to trap carts, but it has never stopped. It is the reason why the old drovers took the high road – some say it is a portal to the underworld.

In the twilight, the ingenuity of the Welsh farmer is being brought into play – he always has a few railway sleepers about his person. The tractor-pull has failed. The two-tractor-pull has failed. The chain has failed, fired like a medieval weapon into a field (but on this occasion, no one is killed). Eventually, with a little modern help from the biggest jack in the world (which came up the valley on a lorry), the sleepers are inserted and the spell is broken.

What happens here repeatedly, although we missed the impact today, is that a right-angle bend, on a 1-in-10-rising-to-1-in-5-hill, arrests the progress of the vehicle. It backs down, thinking it will take the alternate route, the driver turning the wheel clockwise. This is when I am dispatched to scream, 'Left Hand Down!' as I hurtle across the field. Alan has told me to do this – I am faster than him these days but have never understood the meaning of that expression, or indeed 'right hand down a bit'. My father used to employ them all the time and I didn't understand them then. I suppose I should have asked. It is something only men seem to understand, and the reason it took so many attempts for me to pass my driving test – and undoubtedly the reason I cannot back the trailer. Despite my blind use of the mantra, it is always too late. There is a thud as the heavily-loaded grain lorry slumps against the bank. *Sadly*, think our sheep (who have seen it all for hundreds of years) nothing is spilled – carts were much better!

Pedro

When we first arrived in Wales, when our dog was still young and intact and the world was full of willing bitches, Pedro was irrepressible, unrestrainable, clever and devious. He earned us the reputation of the *feckless English who could not even control their own dog*. In this farming area, he should have been shot – you can't have big, powerful dogs just wandering about. The thing was he didn't just wander, he had purpose and inherent cunning. He was never seen anywhere near a sheep, although he regularly crossed their fields. He always kept out of sight, a commando – along the stream or in the ditch. There were no give-away signs of the sheep gathering or running. Ever mindful of the wind direction, they didn't even smell him. He didn't bother them. He had more important things on his mind.

While their bitches wailed in disappointment, angry farmers locked him in barns only to find themselves bedazzled by his escapology. He is always very biddable when caught – it's a fair cop, *chwarae teg!* (fair play in Welsh) he can speak Welsh and do door handles, knobs and latches. One farmer is still scratching his head, like Sherlock Holmes. You see, both doors were locked from the outside and the only window was a good 12 feet from the ground.

It was pure charisma that kept him alive. He would boldly approach the man with the shotgun, wagging his tail as if he'd known him for years. Perhaps he had licked his face one night as he lay drunk in the hedge, when the farmer was trying to get a bit of shut-eye on his way back from a lock-in at the pub. Perhaps the farmer recognised in Pedro his own younger self. They do say that the Welsh (careful) are a passionate race and have their own traditional ways of courting, not dissimilar to Pedro's. Anyway, he survived, and I have written about his adventures elsewhere.

I've never known anyone, human or otherwise, who knows so many people or has so many friends. He's a dog who comes home, after a night out, in the post van (and you know how postmen are supposed to feel about dogs). The postman lets him out at the gate, and he trots home. We benefit indirectly from Pedro's fame. Farmers know a good dog when they see one, even if he is with that damn silly English couple. When introduced to us at chapel they say, 'Oh yes, Pedro's people'.

Pedro has always had a weakness for small creatures, be they baby rabbits which he brings into the house and wants to adopt, or lambs. He bonded with Aby our orphan lamb from the start, insisting on cleaning her after feeds and letting her sleep in his basket. She would follow him about the yard and, one day when he was bored, he taught her how to retrieve a ball, though her mouth isn't the right shape. She can only dribble it triumphantly with her nose, if she gets to it first, but she enjoys the chase immensely.

Whenever one of our other lambs is poorly, he looks at them the way my mother used to look at me if ever I was ill. These lambs remember and never worry about him approaching them again, even though he is a dog. If the sheep hear or see any other dog they are gone in a flash, up the hill to the farthest corner of our land where they form a square, facing outwards, like Napoleonic soldiers, stamping their feet and snorting. Pedro meanwhile will be undercover at a lower level, gathering intelligence about the intruder – he is protective of his flock.

Pedro can't read, but he knows what I am thinking and can smell exactly where I've been. He can eloquently remind me when I forget something in our shared routine. He does this by urgently engaging my attention by gazing through my eyes into my mind then shooting a glance in the direction of the task that I have forgotten (usually one that he enjoys). His body-clock is not, like mine, regulated by an unreliable stomach. It has pin-point accuracy – he monitors the light intensity, correcting for the weather: 'Gosh, is that the time Pedro! You're right. We'd better feed the sheep before it gets dark.' Or, more likely, 'Alright! Alright! Don't nag – I haven't forgotten!' (which, of course, I had).

'What did you say?' asks my husband.

'I am talking to the dog!'

His empathy is as finely tuned as his sense of smell (the dog, that is, not my husband). He knows exactly what I am feeling. Otherwise, he is a pretty generic type of dog. If you had to categorise him, I suppose you would call him a large, short-haired retriever. He is really good at fetching things.

People ask, 'Is he a Welsh sheepdog?' To which he puts his head on one side, looks quizzical and turns, in deference, to me to explain.

'Well, sort of,' I say.

'He certainly understands sheep. A caring-dog, for any of our lambs that are poorly. But he's much more than that. He spots when they are ill and tells us. He's the intrepid mountain-dog and finder-of-the-way-home-dog. Responsible-adult-dog. Always alert, sensor-of-danger-dog. Night-hound, watcher-of-your-back-dog. Ratter, humane catter, licker-up-of-mess-dog. He's very good with languages – he's a polyglot and he's a seeing-in-the-dark-dog, a hearing-for-the-relatively-deaf-dog and a sniffer-dog for the finding-something-dead-job. But most importantly, he is a remembering-dog. Working with the terminally forgetful, reminding me every day of all the 101 forgettable chores on our little farm. Actually, he's more of a general-purpose farm-dog ...' I add, trying to sound sane, then lapse, '... but he's over-qualified. You'd call him a *canis universalis* (is that grammatical?) – a polymath amongst dogs.'

This week, Marley has come to stay while Alison and Dan have a little respite in Spain. Marley is a hound dog, a beagle, ten months old, curious and learning to be good.

The sheep take one look at Marley, from a distance and on a lead, and withdraw to the uppermost margin of the field where they gather, in a defensive formation, ready to stamp their feet and advance as one, heads low, armed for butting, should he approach. They instinctively understand dogs completely, especially young ones.

Pedro looks askance at their reaction. The sheep have ignored him for years, ever since the day when he had been placed to block their way, between the fence and its adjacent hedge. He held his ground, as he had been instructed, like a good dog should, and each of them, in turn, jumped over him.

I am taking the two dogs for a walk to wear out the youngster, but something strange has happened to Pedro. He is in mentor-mode, and having a little helper is as much a trial as a boon. But there is nothing like showing another what you do, to make you realise what a full and interesting life you lead (I found that when trainee doctors came to follow me around to see what country practice involved). Pedro is finding it too.

He shows Marley the ropes: the fences, the hedges, the tracks. He shows him all the holes in the fences where the foxes and the badgers come in. He doesn't have to explain; he just shows him how to sniff them and, by golly, he's got a good nose!

He sees, with his nose, where the badgers get in from the rainforest and what they have done to the pasture. This is the time of the year when the badgers scratch off the turf to feed-up on worms and grubs before winter.

Badger damage is the rootling of the Earth Pig (that's its Welsh name – smells the same), he explains.

This is a run – can you smell a fox? When we find a good bit, we'll roll in it.

He shows him where the toadstools grow, and now we're going home for tea and a nice lie down.

Pedro looks tired. It's been a big responsibility.

Aby's Odyssey

The mud is boiling, and steam rises from the Kennel Field and drifts over flaming puddles. The whole town has turned out, but the flames are so high and the heat so great that 3,000 souls, un-marshalled, stand back in a perfect circle 30 yards from the fire. And the moon looks down from a safe distance.

The little town of Llanidloes is still laid out in a medieval pattern of tightly-packed timber-framed houses within an invisible (long-gone) paling rampart. Not surprisingly then, on the fifth of November, or thereabouts, everyone troops over the bridge to the site of the sheep fair, outside the town – *beyond the pale* (long-gone), safe on the far side of the *Hafren,* the River Severn, for the Bonfire Night celebrations.

What are we celebrating? One suspects that it is nothing much to do with the goings-on of 1605 (it would be uncharacteristic for the local population to be much concerned about events in London). And, looking at the scale of our fire, it is as well that Guy Fawkes was not a Welshman, or the course of history might have been very different.

No, it probably goes back much further.

The fifth of November is the traditional day for turning out the *tups*, putting the rams in with the ewes, and so is really the first day of the sheep-farming year.

Happy New Year!

Autumn colours are appearing, and our sheep are starting to show russet-coloured bottoms. Unlike the flurries of blowing leaves or the spiciness in the damp air, this is not a natural phenomenon – well, not entirely.

Out of a sense of aesthetics, or because the scarlet was out of date or would alarm me, our neighbour has marked the ram that he has lent us to serve our ewes with rustic, mellow autumnal paste, the colour of chestnuts and dying bracken. He (the ram that is) is painted with a thick wadge of this sticky raddle, between his front legs, hidden under his deep, masculine chest, so that, as he discreetly does his job, the ewes turn brown and a flurry of them follow him and swirl around him like a gust of fallen leaves.

Tupping is not what I had expected, with all my human prejudices, when we started farming. We have only one ram – there is no fighting, no rut, no competition for the ewes. The ewes who live together all year, who are related and most of whom were born here, are very welcoming to the ram each year. You can see them shepherding him around the fields, politely showing him the shelter and the stream.

One after the other (and nature is very clever with this), they introduce themselves to him, as each ovulates and comes into season. For a few hours the chosen ewe will follow the ram closely, nuzzling his neck affectionately. He curls his upper lip and sniffs the air to check her out, mounting her tentatively at just the right moment.

Meanwhile, there may be one or two other ewes, about to be ready. They form a guard of honour – bridesmaids following closely, waiting patiently. Something about the chosen ewe, probably her scent, makes the others hold back.

Soon after she has been marked, she retreats a little, although she may follow him for the rest of the day while he gets on with the job in his gentle, quite respectful way.

Next day the ewe is back to normal grazing with her other russet-bottomed sisters. Once all are mellow red (some by then only very pale amber), the ram will settle down too. He will graze with the ewes and still be larger and handsome, but nothing in their behaviour will betray his presence.

But we have a problem.

One of our pretty theave lambs (only six months old, precocious, a foolish virgin) has developed a far-away look in her eyes, and the ram is sniffing ominously at the gate of her field. She and her two half-sisters must be moved out of his amazing nasal range. There are only three as this year most of our lambs were male, but that is another story. The trouble is they have been fed on grass all summer, so do not follow my bucket and have no experience of the treats within.

I need help – come on Aby!

Aby is our secret weapon, our dear old ewe who started life as an orphan in our shower-room and taught me nearly everything I know about sheep. She is my lieutenant, or perhaps more of a sergeant (she's not very academic). She likes to walk around the farm with us, butts the dog for pride of place, and the other sheep will follow her.

Sure enough, she comes running when I call and, having shut the breeding herd up in one of their fields, she rapidly helps me lead the three foolish virgins across the other field, through the woods and up onto the hill, half-a-mile away from danger, on the other side of the trees. It was touch and go when they heard the hunt in the distance. The youngsters pricked their ears and hesitated, but Aby plodded on regardless, and they followed. I suspect she may be a little deaf – she had unmonitored streptomycin as a lamb. Once they are in the top field, they can sit under the hedge and wonder what it was that had made them feel suddenly so strange.

Aby tucks in vigorously to the lush new grass. What an adventure she has had. Today she has led the daughters of the chosen out of peril, defied a mighty king, travelled the length of the known world, faced alien hordes, unperturbed, and tasted the fruits of a promised land – and it's not even lunchtime.

Nothing is new – everything is relative.

Seven years ago (a Biblical period) I probably wouldn't even have noticed. I wouldn't have spotted those white dots on the farthest hill, wouldn't have wondered about the gender of the dots. I wouldn't have rushed up the valley, binoculars in hand.

But next morning, as I approach the boundary of yesterday's place of safety, I need no binoculars to see that our neighbour's tenant has loosed a ram and 20 breeding ewes into the adjacent field – the one with the dodgy fence-posts that falter and play dead when challenged. Through that very fence, this randy tup is sniffing at our precocious theave lamb. She makes me think of Anne Boleyn. She is running daintily up and down by the fence, *baaing* prettily but surprisingly loudly. And each time as she turns, stamping her little feet, the old king on the far side of the fence is transfixed – enchanted. Meanwhile, his ladies gather around his number-one-wife, today's queen, and whisper, looking accusingly at that Boleyn girl. She has a streak of mud on her back. They are questioning her virginity (well, perhaps not).

I run back to the house, opening and shutting various gates on the way, and fetch my magic bucket which has mysteriously been tampered with – it has been used for chain-saw oil, which smells bad (the bucket's power lies in its mystical maltiness).

I chant appropriate expletives as I clean it and refresh its ewe-nut rattle and scent. And a moment later I have my sergeant-at-arms, my witch's peculiar, my familiar. She hurtles to my call – not a swooping, weaving bat, nor an owl, blinking in the light of day, but a clomping old ewe who is trained to my bidding and I to hers. We set off, Aby and I and the dog, armour glinting in the morning light.

Ten minutes later and two of us breathing heavily, we have reached the farthest outpost of our kingdom. It has not yet laid down under the weight of conjugal bliss. I rattle my magic bucket. Anne Boleyn tosses a dismissive glance over her shoulder, but her maids come and have a sniff, and when the sergeant and I turn and walk away, they follow. Anne looks at us, then at the king, then back at us. She is deserted by all her people; she fears nothing more than being alone – she is after all a sheep. It's too much. She turns away from the king and runs down the field, and we all walk steadily back towards the farmstead.

As we draw farther away, the foolish virgins look back. I know that with every step we take, the pull of the king is less, so we keep walking. But then there is a scream on the opposite bank, a hollering and the unmistakable call of a hunting horn (really – not just another escaped metaphor). There is the baying of hounds directly ahead. My foolish virgins stop dead. So does the sergeant-at-arms. They start to turn. I shake the bucket. The sergeant advances towards it (Good old Aby). Our dog, who has been following, aimlessly sniffing for rabbits, suddenly hears the threat and runs to take up a position at the head of our column. As he overtakes the sheep, he turns them and provides a little push and they again follow us.

We stomp on confidently. They follow nervously. I close the last gate behind them as the baying abates. Typical! We haven't heard the hunt for a year – hunting is illegal so they don't exist, but exercising hounds pass through the cutting by our gate twice in one week, just as we are moving timid sheep.

Next, I phone my neighbour: would he like to collect his ram? The job appears to have been done. I am persuaded that every ovine bottom has at least a trace of amber. 'You can keep him until Christmas if you like. I know you always over-feed your stock. It'll do him good – build him up for winter.'

I politely decline.

The chicken fiasco

My very good friend is married to the Chicken Whisperer. Their smallholding is Paradise on Earth for world-weary hens – and some ducks and geese. They live in a woodland glade with a babbling brook and ponds that can be emptied and refilled at the turn of a stopcock. Everywhere is cottage garden and orchards with tumbling verdure and magic, mossy, stone walls sprouting ferns and navelwort. Here is the ultimate gated community with little houses for the various feathered cohorts, groups of birds with special bonds of species, family or long association.

They all return to their own homes at dusk to be locked securely in until dawn, when they are free to potter in the gardens and browse on nature's bounty, or feed from the bowls of delicious and varied porridges that my friend prepares under instruction from the Whisperer and which cater for their special dietary needs.

A few weeks ago, a hen disappeared. Searches were instituted. The ground was scanned for feathers. Every nook and cranny was probed – no hen was found. Security was reviewed – electric fences and nocturnal patrols were discussed. Then she reappeared!

She was not alone – behind her marched seven chicks, brooded in secret and now displayed to the world. But every night, just before dusk, they disappear again.

My friend and her husband hide in the bushes, peep around trees and skulk in the lane, but cannot find their hiding place.

Every morning in trepidation they count the chicks. Every morning there are seven – now almost as big as their mother. The Whisperer and his wife are wan with sleepless anxiety about this stubborn mother hen and her at-risk offspring, out in the night to be smelled out by a fox.

'What they need is a new house – their own place!' Timber is purchased, and roofing felt and dowelling for perches, door furniture and hundreds more nails and screws than were actually needed (that's hardware retail for you these days). Digging and levelling, sawing and hammering ensue. It takes a couple of weeks in the rain and wind, dodging falling branches as Hurricane Ophelia comes and goes. Still, every morning seven chicks appear and march in step past the work-in-progress.

Then humane traps are constructed and baited deliciously (these chicks are not stupid) and the Whisperer knows that it has to be all or nothing – mother hen and every single chick or no-one. To leave one or two chicks alone in the wild night is unthinkable. Catching them all takes enormous concentration and time (two whole days) and lots and lots of treats. But Bingo! They are all caught and decanted into their beautiful new home. They are shut in for two days and two nights (a lot in chicken-time). 'That should be enough', says the

Whisperer, confident that now they will return each night to their secure and luxurious new accommodation.

However, they do not. On the third day, at dusk, their coop is empty – no mother hen, no chicks.

But hey, what's this? My friend is showing me a photograph. Upwardly mobile chickens! Not very clear, but they are all up in one of the tallest trees. That's right, you can see the top of a telegraph pole which gives away their altitude, but the falling leaves have denuded their cover.

Arboreal chickens – what next?

~

The fallen leaves are rushing about the field like demented mice, agitated by the wind, running and leaping then taking off in a murmuration, swirling about joyfully then crashing against the window, driven by the heartless wind.

The air roars and tumbles and swats around the house, and crowds of raindrops suddenly applaud – driven from different directions, in thrall to the wind, their little bodies clapping against the walls, the roof, the glass.

The trees on the horizon rock with uneasy mirth, the firs more nervous than the now-bare oaks that stand complacent and let the gale comb through their nakedness. The lone pine thrashes like a wet sail in the hands of a novice, tested by the flailing gale.

It is the mating season, including for trampolines. A male bowls across the hillsides looking for a mate, tumbles down the bank and leaps over the hedge. He bounds down the slope then soars on a powerful gust, trailing his long-netted plumage as he hurls himself down the valley.

Storm warning in Wales – 150mm rainfall this weekend.

The wind moans high above our house – we are in the lee of the hill. It was built where the sheep sheltered: in the 1840s they noticed those things. I notice the tall trees at the end of the house flailing about in the turbulence of the mounting storm, but Alan will not sleep in the spare room at the other end of the house, out of the reach of falling branches.

In mitigation for my cowardice, I'll tell you that two branches of Douglas Fir *do* fall, crushing a steel hurdle but missing the old and rotting chicken coop. It has been gradually sinking into the mud in the last week as the rainfall has reached 109mm. It's been a dry summer and autumn, ever since I started measuring the rainfall: only 718mm since April (believe me, that's not much for here), so we know we are in for a deluge – a couple of metres at least.

In the village, down the valley, a man is startled by an unfamiliar shadow. He looks up into the great oak tree on the edge of his yard to see a skeleton hanging, draped in a black cape – it is the spent trampoline, like a giant dead crane fly.

In Britain we do something very strange in autumn. We turn back our clocks one hour, to give us more daylight in the morning and less in the evening (is that right?). It started in the War when someone decided to put the clocks forward in the spring to give us long, balmy summer evenings in which to 'Dig for Britain' and increase agricultural production. Ever since we have been moving the clocks back and forth and generally confusing ourselves.

Last weekend was the time designated to move the clocks back (we are all supposed to do it at the same time). Not that the hours of daylight are impressed – they have continued to dwindle along their inevitable celestial way, getting shorter and shorter, leaving less and less time for the farmer's chores. And we have to cope with the disturbance in our routine wrought by the hour change. Waking too early, hungry at all the wrong times, confused animals, missed liaisons, getting to the dentist at the wrong time and general discombobulation.

I blame this for the chicken incident.

Last weekend we went to Rutland for a wedding and Alan (who is thus culpable) noticed a sign saying: *Point of lay chickens for sale.*

Now, anyone who knows anything about chickens knows that they stop laying in winter, which in the Northern Hemisphere comes shortly after autumn. The purchase of chickens at the point of lay in autumn is pointless. They (the chickens) will quite likely be eaten by hungry predators during the long, dark, egg-less winter months, and you will never see the fruits of your investment.

Note to self: buy chickens in spring. But sometimes one just wants to do something extraordinary.

If you were here in person, I would introduce you to our new chickens ...

Those astute amongst you would notice something strange about two of them. They are brown and smaller and constituted in a different mould from the other two. More low-slung – like a sports car compared to the others, which are grey saloon cars. Their bills are different, not so sharp, broader and flatter. You've got it – they are ducks!

We did not lose our powers of reason entirely in Rutland. We noticed the huge pile of eggs on a box in the corner of the strangely muddy yard: duck eggs.

'Oh yes,' said the lady, 'very good layers, our ducks, and they go on laying throughout the winter'.

So instead of four hens, we drove home with two grey hens (a Speckled and a Bluebell) and two Khaki Campbell ducks, shut in the boot of our camper van.

We learned something new almost immediately, as strange smells emanated from the rear of the vehicle unsettling the dog. Ducks, unlike chickens, do not switch off automatically when placed in the dark; they can see in the dark and they can squeeze through surprisingly small holes.

When we got home, we had two chickens but no obvious ducks. After the removal of several panels, the mattress and parts of the bed, we found them, gone to ground, between the water tank and the chemical toilet (obviously not liking to poo on the plastic sheets and newspaper that we had put down for them in the boot).

This week, while I shampoo the camper van carpet, Alan (partly culpable) has been constructing a new pen and coop for the production of the most expensive eggs known to man.

Meanwhile, I have a nagging worry about the ducks. Raised commercially from day-old chicks in a yard with only shallow trays of water and small puddles, and without the benefit of proper parenting, we may well have to teach them to swim. Life is full of new challenges ...

They have laid nine eggs in just under a week and are very meticulous with their ablutions, taking a good hour to do what Granny called 'a stand-up wash' every morning. Then they flap around drying their wings.

Last night, late, we arrived home from a week's holiday in Cornwall. So, only now, it is safe to mention the feverish activity that preceded our departure (we do not believe in tempting fate).

Now we are back home, I can risk telling you that feverish

predator-proofing preceded our departure. We struggled in the torrential rain to make the poultry fox-proof and badger-proof before entrusting them to a friend to tend daily. We erected eight-foot-high double-fences with embedded roof slates buried and wired around the base, and heavy scaffolding poles also fixed at the base to discourage tunnelling. Alan vetoed the purchase of electric fencing (God forbid: that is in plan 'B'). Phase Two will encompass the netting roof which will be necessary when we have chicks, to keep out the magpies and buzzards (but I haven't mentioned this to Alan yet).

The cats had not left home this time and were indoors waiting for us on our return (the one dubious benefit of heavy rain).

I immediately donned my wellies and went out to shine the repaired torch through the chicken-wire window of the new coop. The chickens were on their perch, one with its head cocked quizzically to one side and a brown egg smashed on the floor beneath her – 'point of lay' but still hasn't quite got the hang of it. Never mind, she'll stop soon.

The ducks were also on the floor of the coop, carefully preening the last of the day's mud from their feathers in a scene that reminded me of our bathroom on a Friday evening when our five children, then teenagers, were still at home.

Outside the coop, in the new predator-proof poultry run, something strange has occurred. A 25-foot square (that's 625 square feet) enclosure of pasture has undergone some sort of cold fusion. There is, it seems, a complication of *keeping predators out* and it is *keeping poultry (especially ducks) in.* In one week, two ducks have carefully liquidised the chicken run.

They meticulously probe the soft soil for grubs and wriggly things, repeatedly washing their bills in any standing water they can find. Puddly land becomes a morass in no time (they are very conscientious).

~

This morning, bright and early, I count the sheep huddled by the fence and find one too many – that's odd.

One stray sheep has joined ours and is trying very hard to blend in – they do so hate to be alone.

In this wet weather sometimes, in the dips, the tension in the wire fences lifts the fence posts right out of the ground, and some of the ewes are quite clever at encouraging this process, particularly if there is nice grass on the other side. This seems to account for our new ewe – she is returned.

The exceptional rains continue, the ground is saturated, the reservoirs are overflowing and the rivers are in flood. It's worse further north and it's bad enough here.

Last week, unusually for December, I saw a Dor Beetle on the path – *moving to higher ground*, I thought, *ahead of the flood.* Just in time, as it happened, before water started to spew out of its burrow. In the valley bottoms, the water table is higher than the ground so that the track feels like a water-bed as you walk along, and molehills erupt with water, like volcanoes. If you stab the ground with a stick it may spurt at you.

Water is running everywhere. Waterfalls appearing where they do not normally belong.

At dusk this evening the chickens put themselves to bed, but the two ducks are nowhere to be seen. They aren't in the sodden field, nor in the yard; not under the truck and not in the barn or in the road which has turned into a torrent.

The torch won't work and the hurricane lamp blows out.

But somewhere above the storm there is a distant quacking. The two ducks have strayed into the wetland (well, it's all wet at the moment) and become separated and are calling to each other over the stream. As I approach through the aspen and alder, one panics and tries to cross the raging stream (remember, their wings are clipped). Next minute she is in the churning, muddy water, whizzing downstream, spinning and flapping, quacking and squawking. I am downstream of her, so holding onto an overhanging tree, I manage to lunge at her as she approaches, and flip her unceremoniously onto the muddy bank where she disappears into a holly thicket before I land, *splat*, where she had just been. Traumatised (both of us), I carry her home over the bridge, and reunite her with her friend, who comes running to meet us.

Tomorrow, I think they'd better stay in their new enclosure. It's too rough for ducks.

The febrile planet

When one of my children got a fever, she would rush about randomly picking things up and putting them down and talking too much (quite sweet but scary). We knew that if we didn't cool her down, she would have a convulsion. The more energy you put into a system, the faster it goes and the more unstable it becomes (even a little girl).

The weather has been like that. As things warm up, the system speeds up. The winds whizz around the globe picking up more moisture and dumping it in ever increasing amounts. The winds blow faster – the whole thing gets unstable (from a human point of view).

In fact, I suppose, Nature is doing what Nature does best: she is resisting change, using all that extra energy to blow and suck and push and pull. To evaporate the seas and to lift the sodden air and swirl it around to generate static electricity and throw lightning bolts around the heavens, melting telephone lines in Wales and flooding the low-lying areas of most of Britain. The media doesn't even notice that parts of the United Kingdom have dropped from sight. At first just soggy, then gone, submerged.

It's warm and the grass is growing, but it just won't stop raining. I know it's Wales, but it has poured unremittingly for three weeks. My rainwater gauge records 260mm, and 60 of those in the last 24 hours. As I go out for some milk, I have a Dr Gloucester moment. Splashing through puddles in my little car, it suddenly feels as if I am driving through treacle and the outside world disappears under the wave that envelops the windscreen. Where is the road?

It reappears only to disappear again almost immediately as I realise that I am wearing the wrong vehicle. I go home and change. Alan puts on his red woolly hat (a sign that he means business), and we set off together in our truck to intrepidly go and be amazed by the awesome power of water.

Everywhere sheep were damp and disgruntled.

Penforthlas is a village clinging to a windward Cambrian hillside, and extending down into the beautiful, very green, Clywedog Valley. Its old-English name was an invitation to *Stay-a-little* and have some tea and a chat, but later corrupted by immigrants who couldn't tolerate the wind and the rain, and called it *Staylittle* before they left. Here the water is rising.

By the Clywedog Reservoir, whose job it is to regulate the flow in the River Severn, men from the Water Authority stand by the dam and watch silently. Their body language is not encouraging.

The water in the reservoir is creeping up the fingers that extend into the hills on either side, and the river that feeds it, normally running tidily within its banks, has overflowed and is trying to reach the lakes that have appeared in the adjacent pasture. Soon it will be an inland sea, lapping around a footbridge that now goes nowhere. The bridge practices to be a landing stage and the chapel is bracing itself to become an island.

The rivers *Gwy* (Wye) and *Hafren* (Severn) both start within about a mile of each other on a hill just up the road from here. Llangurig is the first town on the Wye and Llanidloes is the first on the Severn. Neither of them is really a town at all. Llanidloes probably has fewer than 2,000 people in winter, but was once quite thriving and industrial – certainly industrious. Llangurig is much smaller but on an important junction where folk turn right on the way to Aberystwyth, so it is on all the maps and signposts, which has rather gone to its head.

Both these towns are very near the sources of their rivers which go on down their respective valleys gathering volume and momentum. We have never seen them rage so much and so soon, and so we fear for the communities downstream.

Today, while I stand taking photos of the river by the Old Mill in Llanidloes, I meet the architect who is looking anxiously at the mill flats, converted about ten years ago. He comes up to the bridge that overlooks the mill whenever there is particularly heavy rain, to check the level of the river. He tells me that he has never known the 12-foot arch to be submerged completely before. It normally stands with its feet quite dry on the big rocks that rise above the splashing stream as it picks its way between the boulders of the riverbed, its waters dividing and reuniting, jostling and gurgling past. Sometimes here we see a dipper, a plucky bird not much bigger than a blackbird, wading upstream just below the surface, looking for things to eat amongst the pebbles.

Today the waters are brown and heavy with sediment, moving as one body, surging past, pressing against the bridge on which we stand. We move across to the other bank.

It's all a bit worrying. The rain has stopped now, but everywhere roars with draining water.

In Llangurig, the Wye is only about six miles from its source. Not such a huge catchment really, but today the back road to Rhayader suddenly disappears into frighteningly fast-moving flood-water, an inland sea with perilous tides. It is 400 yards to the hills on the far side of the valley, and the sliding brown water stretches the whole way across. It will have to drain all the way to the sea, collecting water as it goes.

It will churn along the stretch from Builth Wells, with great breakers carrying away the smaller trees that stand too close to the now-submerged banks. Through Hereford, Ross and Monmouth, all this water and more will flow. It will heave through the Forest of Dean, swerving in a great semi-circle around the ruins of Tintern Abbey and on to the walls of Chepstow, then eventually join the waters of the Severn, and on to the Bristol Channel, carrying tree trunks and café tables, dead sheep and plastic bottles, all hurtling towards the sea to meet the tide.

The power of all this water is stupefying.

~

For all this, Alan and I relish the extreme changeability of the climate in which we find ourselves (blame it on our Pennine childhoods).

He was born in Stacksteads, Bacup, during the war, in an industrial area with cotton mills, mines and smoking chimneys that spewed black smoke that washed down in the acid rain to erode the blackened sandstone buildings built at the peak of the Industrial Revolution. The place vibrated with the thuds of steam engines. The foggy mornings were rent by the whistles of steam locomotives as they roared down the Rosendale Valley, pulling a pall of sooty steam that swirled around Alan, as his mother pushed his pram under the railway bridge, then along the bank of the Irwel river which was turned blue by the dye factory upstream – last week it was yellow. They were visiting his grandmother in her tiny house built half underground on the side of this capricious waterway. It was dark and damp. By the time he was a year old they could see the glow of the fires from the heavy bombing of Manchester.

I was born in the next big valley, not that we were to meet for another 50 years. I spent my early childhood in a farming village (no surprise there), overlooking Darwen, a Lancashire mill town where Dad worked as a lowly manager in the post-war chemical industry.

Looking back, I think my mother might have been depressed when I was small. She had had two babies in quick succession. I was the second – the first had died at birth. She had moved up to Lancashire where my Dad had got a job. Like many couples, they had lived apart throughout the war, and she found Lancashire, and probably my Dad, very strange and different. My father was still extremely jumpy and had terrible nightmares 'because of the war'. And, to top it all he had an affair. She tried to leave him but was sent back by her father. Thereafter she stayed at home with the secret shame of her husband's infidelity, and beavered away being the best housewife she could be. I was oblivious to all this at the time, until the day before she died – I wish she had told me sooner. For that short period of my childhood, other people were my salvation. Lancashire was my salvation. Friendly, tolerant people, amused by the little girl who, given the correct footwear – wellies or clogs – stomped behind the local farmer, almost from the time she could walk.

It was 1952. That was me, following the grown-up I called Uncle Harry. The first word I uttered was evidently 'Baack!': the sound I heard as he backed his old carthorse into the shafts of the cart that delivered the milk. I lay on the other side of the wall of their farmyard in my pram, looking up at the leaves moving in the wind. It was not my mother's face I saw, cooing and smiling in at me, but the waving bough of the rowan tree. Is that why now I feel such a surge of love when I look up in the woods?

Harry Bland and his wife Eve farmed on the edge of this northern mill town, in a place called Sunnyhurst. They kept chickens and geese for eggs and for the pot, and cows for milk – probably never more than 17. He farmed sustainably, with low overheads and by selling his milk only to those to whom an old horse could walk each day.

That is not to say that Harry was not progressive – his herd was tuberculin tested before anyone else's. My Dad helped him fix headlights to his Massey Ferguson tractor so that he could make hay after dark or before dawn if rain was threatening. Then everyone would help. This was just after the

Second World War, and, from the time I could walk, I was there too – or so it seems, looking back. I collected the eggs, helped feed the cows, listening to their lowing and feeling their steamy breath on my bare legs. I filled nets with hay and hung them up on the bars of their stalls. Sometimes I stood on a stool in the wet dairy, washing milk bottles with my hands in gloriously warm water (I still do my best thinking in a hot bath).

In the next room, stood in my wellies, I could watch Mrs Bland fill the bottles, dried in the hot cabinet, with milk running from the cooler which hung on the wall. It was like a radiator but with cold water circulating through it and the milk running down from a reservoir on top, filled from the heavy milking machines. These were electric. Harry would carry them in from the shippen, unscrew the pipes and uddery bits to wash, then haul up the heavy base to pour the fresh steaming milk into the cooler. I don't remember the pasteuriser.

As a treat I would sometimes do some bottle-topping with a hand-held press that moulded silver tops onto the modern bottles that went into a crate to be delivered. There were still some old bottles with wide tops that were sealed with cardboard discs (too fiddly for me). Most milk was sold without being bottled, ladled from the churn on the back of the cart into *kits* – metal cans with lids, presented by the customers for Harry to fill as he walked the old horse about his round.

If I got bored, I would go round the back, scuffing my welly boot heels on the wet concrete to make that farmer's noise when I walked. I'd go and help Harry sweep the manure and urine from the channel behind the cows with a heavy, bristly broom and enjoy the aromatic odour (I wonder what that means about my animal nature).

I don't suppose I was very useful, but with my back braced against the wall of the barn and my arms extended, I could hold two flapping chickens by their legs after Harry had wrung their necks. If laid on the ground, they would get bruised. My job was to hold them up until they stopped flapping, while Harry got on with the next two. This was a great responsibility of which I was very proud. Sixty years later I surprise myself when confronted by a severely injured pheasant. Like Queen Elizabeth II, I know exactly what to do.

After milking we would walk the cows back to the meadow with Rover, the dog with his matted coat and the orange tummy (stained with the slurry). Harry and I would walk side by side, and he would tell me which bird call was the peewit and which the curlew. We would stop to watch the water gushing out of the culvert and tumbling down the gully. This made a magic, secret sound that you could always hear as you passed by, but whose source you could never see unless you hoisted yourself up to look over the high dry-stone wall at just the right place.

When I was not very much older (for we left there when I was five), I would sometimes go with Harry to collect the milk-money on a Saturday when he went a bit later. I would sit in the cart and talk to the ladies in their printed-cotton pinnies, hair tied in cotton triangles, most with curlers in. Terraces of houses covered the hillside like the ridges and furrows of a ploughed landscape. The furrows between the backs of the terraces were strung with dancing, drying linen. Uninhibited, jolly (probably bawdy) chatter, from groups of women, followed the cart while little children, some barefoot, played along the backs.

Once on a road of soot-black houses, I was alone in the

milk cart when a passing car – there weren't many in those days – backfired and the horse took fright and bolted. It tore along the cobbled street, its mane blowing in the wind, and I became aware of Harry in his flat cap, his brown cotton coat flapping, running for all he was worth next to us and grabbing at the reins to steady the poor creature. That was the first time I saw real fear on a man's face. He caught the horse before we got to the steep hill down to the main road – Harry saved me.

I think he did that in more ways than one.

I spent a lot of time in Harry's cowshed. I felt really at home with those steaming beasts and they with me. After all, they were dairy cows and their calves were taken from them at birth, and I was a tiny creature struggling up and down the gallery behind their stalls, carrying a galvanised bucket of oats, half as big as me. All amidst that yearning maternal love, they nuzzled their pig-tailed attendant – but not all of them.

I soon learned that personality was not solely a human trait. There were cows with great empathy, kind cows and ones you had to watch – grumpy, irascible cows who would swing their heads and knock me flying. But the more you knew them, the more you understood.

I was allowed to name cows (a great privilege), and I duly named them according to their personalities, after people they reminded me of.

There was Eve, the scrawny, vociferous brown cow who mooed her demands at me. She never, ever knocked me over, but watched me carefully, following me with her eyes. Eve was the name of the farmer's wife.

Grace was a slim and youthful, clean and crisply-marked

black-and-white cow with sad, moist, longing eyes. She was named after my mother.

My names for the animals used to amuse the farmer and his lad. I liked them, and I liked making them laugh – they were my friends.

As I grew up, I found that the world my parents inhabited – for they were not farmers, and they were not Lancastrian – was a lot more complicated, and that people had to be judged according to peculiar criteria which made no sense at all. You were not supposed to choose your friends from the people you liked, but from designated groups selected by age and gender and social class. More particularly, by the way they talked (and in rural Lancashire, the Queen's English was hard to come by), where they lived, and by something called *table-manners* (unless you were French) – I never could quite swallow this.

As I have got older, I find (as with many things, for I am unattractively opinionated) that I was right all along, and when people talk about the *University of Life*, I think one should attend the elementary school of the farmyard, where what counts is not species (you can't help your species), but character.

Alan tells me that when he was little, he was sent to spend the holidays with his aunt, uncle and cousin in Gainsborough, Lincolnshire, where his uncle, a dapper man, was market manager. His auntie would rush out into the garden to tell Alan not to be so loud – he soon realised that it was not the volume that jarred, but the Rossendale accent!

My mother sent me, screaming and kicking, to elocution lessons when we moved south, to eradicate my Lancastrian vowels and readjust my identity – she knew about linguistic prejudice. That's why, at the pub in Wales, they call me *Lady Diana!*

Alan worked it out for himself, with the help of his auntie perhaps. No one can ever quite work out where he comes from by the way he speaks: somewhere around here probably, they will conclude.

Wherever *here* is.

~

Alan brings me a cup of coffee in the morning. Sleep lingers – I am not good in the mornings.

'Good morning, Sweetheart,' he says.

'I've lost an umlaut!'

'It'll be under your pillow.'

'I'm worried, it might have fallen into the Diphthong.'

'It'll be back in Lancashire then.'

'I'll never find it there – the ground is littered with aitches. Gareth says he doesn't understand a word I write, but you do, don't you?'

'Go back to sleep, or you'll never find it. Have you looked in the Co-op on Duckworth Street?'

'How clever of you – I remember now ...'

Lambing without Spring

The Welsh Assembly Government has written to all its farmers offering help and support in their recent, often devastating, difficulties.

This year starts as one of those when Life is determined

to demonstrate that it is *no laughing matter.* Survival should not be taken for granted. There has been a gradual erosion in confidence over the previous months. It is the wettest year anyone has known – confidence, like rock-salt blocks, erodes rapidly in heavy, unremitting rainfall. The buoyant farmers of Wales no longer bob on the surface of life's challenges – they are water-logged, submerged, just below the surface, along with the frozen frog spawn in the lagoons along the flooded farm tracks.

Just as the early lambs arrive in the fields around Kerry, and people remark that the daffodils are late this year, the blizzards start. There is a Welsh saying: *eira bach, eira mawr*, which means literally, *little snow, big snow* (typical of the non-specific nature of the language). An Englishman would say *fine snow, heavy snow.* You get the drift – there are blizzards of fine, sharp ice crystals, hardly snow at all, but *eira bach, eira mawr* – deep-blown drifts form over the ridges and behind the hedges where this ice-laden east wind funnels through the holes, making bizarre sculptures 12 feet high – frozen serpents rising from frozen, breaking waves in a frozen sea.

Arctic air hits the hills of Wales where the hill-farms remain, and where shepherds still tend the flocks which lamb, not in sheds, but in the fields, in the lee of the ridges and behind hedges. This year the lambs are born into the eye of a *cryoclastic* storm. Many die within minutes of the searing, bitter cold, or more slowly in the torpor of hypothermia in which they sleep to death without the life-giving warmth and calories of their mother's milk. The ewes are starving, trapped by drifts and debilitated by the poor grass and silage produced on all those rainy days last summer. And their little bodies are covered by falling snow.

Those farmers with sheds gather as many as they can, providing what shelter they're able. The ones whose sheep are tamed a little, are accustomed to run to the rattle of the pail or the rustle of the feed-bag, thank God, as the ewes run down the slope through the thick snow and down to the yards before the tracks and gates are blocked. But how many have left lambs on the hill? Losses are heavy and the wind continues to blow from the east, dry and cold, sweeping the snow into gullies. One week into April, after Easter, the mountain road is still closed by snow, but the dreaded floods are stopped at source. The snow does not melt. It slowly evaporates in the wind which dries and scalds the grass. The roads are left dry and stained white by salt.

Still stinging from the insult of last year's lamb prices (which were pitifully low following a good lambing year), one might think: *at least what lamb there is should sell well this year.* Those reared indoors until the spring arrives should save the day. In 1947, it was only the sale of sheep skin, preserved on the frozen bodies until the thaw, that kept some farms afloat. But Nature is not so easily appeased this year. Abortion rates are high, and sheep, unlike humans, do not usually abort in the first trimester of pregnancy. The ewe carries the *conceptus*, investing it with all her strength and nutrition, until the last third of the pregnancy – often to the last two weeks when the main causes of contagious abortion strike. This year it spreads through their crowded sheds. Even the provident farmer, who has vaccinated against abortion, is hit by the scourge of *vibriosis*. The lambs abort in the sheds and many of the ewes die.

Older farmers and those with long memories lamb later. And, sure enough, the spring does eventually come. The tiny, yellow stars of celandine appear under the hedges, between the crescents of still unmelted snow which, from a distance, looks like new-born lambs, bright in the sunlight.

Farmers are now turning out ewes and lambs onto the shocked fields. There is a flush of new green on the wind-scalded hillside, and lambs, unaccustomed to the brightness, take tentative steps into the open.

The thing is that shepherding, though part of the farming industry, is about the relationship between farmer and stock. There is absolute mutual dependence. The shepherd loves his animals. People find this difficult to understand – farmers find it difficult to understand. A year like this takes its toll.

~

But what of previous years? Wales is still a place of oral history, of cups of tea and a slice of buttered *bara brith*, and a sit down in the warm kitchen.

And of remembering.

I'm too young to remember the Big Freeze of 1947 – only just, but my mother was there, living on the Pennines in Lancashire, and she told of walking up on the moors after the worst of the snowfall where it blanketed the ground. She followed the contour of the hill and the walking was surprisingly firm and easy. Suddenly the path she walked was gone from under her advancing foot and she plummeted downwards. *Mine shaft!* she thought. But no. She had been walking along the top of a 5ft dry stone wall and she had simply reached the corner of the field and fallen off the wall. Fortunately, my father, a much more cautious individual, was following and managed to heave her up onto the wall for a more reticent and wobbly return journey.

She also recalled men pulling ropes, hauling a coffin on a sled up the lane, past their cottage, and around the hillside to the village of Tockholes which was cut off by the snow. The as-yet-unused coffin had been filled with loaves of bread to sustain the isolated community. The deceased could rest comfortably and safely in his coffin by his back door until the thaw, when a lorry could collect him, bound for the undertaker in town.

During the Big Freeze of 1947, farms and communities here in Wales, that were cut off by the snow for weeks, had no way of communicating with each other or the outside world. Those with a wind-up radio might pick up a BBC Welsh bulletin or the weather forecast (not very reliable in those days). On 4 March, 1947, people were out tending their sheep, getting in wood, doing what they had to, despite the cold weather. An old man had died in Llys Maldwyn in Caersws, the workhouse until 1930 and thereafter still a place of social care. His body was to be returned to his friends in Llawr-y-glyn for burial locally, where he had lived and worshipped. A grown-up daughter of this farm was dispatched, on foot of course, to arrange his funeral over the hill, at one of the chapels in *Penfforthlas,* Staylittle.

She had not ventured far up onto the moor when the snow came – a blinding, driven, freezing blizzard. The family waited, thinking she would turn back, but she did not reappear. Jack, one of the brothers from Cwm Lowe, a near-neighbour, bravely set out to try to find her, though he was soon driven back by the blizzard. No one came from that direction that day, nor the next day, nor the next. It was four days before the weather abated and her family was able to venture out. As they looked up the dell towards the moor, they saw a hunched and muffled figure trudging along the ridge through thick and powdery snow. They couldn't make out whether it was male or female. Approaching their land along the steep bank, the figure missed its footing and tumbled down the hillside, over and over, all arms and legs, in a flurry of snow like a great rolling snowball. It was the missing daughter, returned in style!

She had been forced to take shelter at the first farm she came to. She was given dry, warm clothes and helped out of her own cold, wet things, her long skirt was frozen solid so that it stood by itself by the fire. There was no way to get news that she was safe to her family. Telephone lines were creeping, like fungal hyphae, from the towns to the villages, but they had not yet reached the isolated farmsteads of this area. The telephone did not arrive until the early 1960s – about the same time as the electricity.

Other folk were stranded in 1947, taking shelter where they could. A young man from Staylittle went up to Dylife to feed his sheep. The weather was rough. Before going home, he stopped for a cup of tea in a farm in Dylife, probably hoping the wind would drop. The wind did not. It was so high that sheep were being blown across the yard and hitting the wall with such force that they were killed. There they lay, up against the wall in the midst of a swirling tempest. Despite this, the young man set off for home. The thaw was coming, and the snow was only just above his wellingtons, but wet. He got down to Staylittle, but cresting the ridge on the far side of the valley, where he would normally be able to see the familiar buildings of his own farm, he could not even stand, so hard was the wind blowing. It drove the sleet and rain horizontally against him so that his coat, heavy with water, thrashed about him. He was buffeted back to a farm where he found a lady milking. She took him in and lent him dry clothes, but it was the early hours of the morning before the wind had dropped enough for him to cross the valley to his home and his worried parents.

Many sheep were lost that year, buried alive in huge drifts. Farmers struggled to dig out those they could. Dogs that could sniff out a buried sheep earned their keep that winter. And, with their owners, would help their neighbours search for their stock. One ewe was dug out alive after five weeks! Even when the thaw came, more were lost. In the high winds, sheep were blown into the swollen river and drowned.

Losses were devastating: farmers lost hundreds of sheep. Even if they were not buried in the drifts, there was not enough hay to keep them going for so long, and no way of getting more. Not for a long time had the bonds of community, always great in this unforgiving landscape, been greater.

~

The *craws* of crows and jackdaws puncture the sound-scape of hills and meadows. Aerial battle resounds. Broadsides ricochet in the pale sky above the passerine chit-chat and base-line *baas* of our valley.

A new chord rises – the dog points, ears pricked, and sniffs.

A strange, still wind?

Rumble of some terrible upheaval?

Discord?

Birds pause. Listen!

It rises from the supernatural, from our eternal underworld, louder, voices more distinct – a celestial choir.

Some 25,000 souls look up from grazing to acknowledge their lord, each with a different note from the human range – angel range.

Hear the crescendo from three miles away, each note swelling with excitement, a wave of emotion to touch the very core. Now the melody is with the bass – diesel baritone – and percussion over the cattle grid.

Robert is driving his gator up the road that runs along the floor of our valley. He is coming to feed his sheep, starting at the farthest point from home and, as he passes the fields along his way, the hungry ewes stampede to their gates and stand to attention. As he passes, they scream that he has missed them. Only as he returns, emptying sacks as he goes, does peace return to the valley. Listen very carefully now and you might hear the low moan of myriad munches.

The ewes and lambs have at last been turned out onto the pasture, which is still shocked from our terrible winter, not lush with new grass as it should be.

As I feed the ewes, the sun comes out, illuminating two dandelions by the shed, and they lift my spirits (we've had a difficult few months).

And in the old sink by the door, peeping out from last year's leaf litter, like a prayer, are primulas, bright and new.

The sort my mother used to plant.

Living on the edge

We take nothing for granted. Our water, which is sometimes golden, comes from a well half-a-mile from the house, up a steep hill. In the recent, relatively dry weather, we have cut a track to it so that when filters block or tadpoles appear in the bath, we can reach it without angina, on our four-wheel-drive, but not-so-trusty, quad-bike. We don't take that for granted either – last winter it also developed a blockage. Outside it was seriously sub-zero for months, so that one's ungloved fingers froze to the metal gate latches and to the carburettor of the quad-bike. I took the feed to our 30-odd sheep every day on foot – well mainly on foot, quite often on bottom. In fact, I soon rediscovered the joy of sliding back to the cottage on the empty feed-bag, pursued by an excited and snowy-nosed dog, and watched by puzzled sheep.

Anyway, having beaten a track to the well, we can't resist a little tinker. Like all our neighbours (some of whom are still quite young), we are preoccupied with our *flow*. The filter is clean, and the well is brimming with crystal-clear water. But Alan is convinced that we have a stricture, some sort of partial blockage somewhere between the well and the house. In mitigation, my husband has a family history of municipal plumbing. His father looked after the water supply for all the mills and terraces of Bacup. A man who, within his head, had the only map of the town's water supply, which he kept up to date by meticulous examination of the innumerable underground waterways and any muddy hole that was dug within the borough. Alan now, having finished knocking down the bracken on the track with his lover, the ancient Hitachi six-ton digger, decides they need to dig a hole.

The chances of him finding the pipe leading to the house, which is the object of his quest, seem unlikely. But within half-a-day and only three holes, he has succeeded, and the flow seems better than we dared hope for, as it spouts into the air from the accidentally-severed pipe.

Forgetting the mended quad-bike, he runs all the way to the well to turn off the stopcock, which breaks off in his hand. Already in a state of collapse, he is not thinking straight. He throws open the well (startling the frog), and pulls out the top end of the pipe so that the water is saved.

The one thing you need to know about wells is that you do not need air in the pipes. And the one thing that everyone knows about Wales is that there is never any shortage of water. To emphasise this point, it starts to rain. The track, 45 degrees in parts, becomes slippery. Temporary repairs are made to the severed pipe (did I mention that it is 5.30pm on the Saturday before a Bank Holiday, or that we have fastidious guests arriving at any moment?) There is now *no flow* – no water at all to the house.

However, there is always help at hand and there are friends, one with his pump and the other with his generator. And there are always bits of baler twine and jubilee clips and me to push and slither. Generators are heavy and pumps are playful. I am sceptical.

The game is to pump enough water from the top, fast enough to drive out the air from the open tap in the house. The pump, enlivened by the grumpy generator that has not enjoyed being hauled up such a dangerous track, blows off its jubilee-clip and soaks the elderly but enthusiastic workforce. Three times the joint explodes, but on the last occasion its owner takes it in-hand, thrusting his thumb over the end of the pipe, which nearly disappears up it: it is sucking! The air has gone. The syphon is re-established. The end is shoved into the well, thumb still attached. As he pulls out his thumb, there is a satisfying *slurp*. And the frog that has been watching from a vantage point, half-submerged near the edge of the well, starts to drift precariously towards the sucking pipe. There is shouting and fumbling and splashing as the flailing amphibian is rescued and the filter refitted.

All the bits of kit have now been returned to the neighbours (cleaned), and all the temporary modifications undone. The pipe has been permanently repaired and hardly leaks at all, and a new stopcock has been ordered and the flow ... well, the flow is marvellous, at least for a few days, the pump having blown out all sorts of gubbins. But, you know, in recent days it's been dwindling a bit.

~

Our digger is something else that seems to have a mind of its own. Since it accidentally cut off the water supply to the house, it has been moody, often refusing to start. Alan and his chums stand around it, scratching their heads and talking solenoids.

The one time Alan's two grown-up sons sneak up to where we are trying to dig a pond, they manage to start it, but then get hopelessly stuck in the mud. They are frantically trying to extricate their father's much-loved Hitachi from a boggy burial. I run around proffering advice and am nearly buried alive as I heave a tree trunk into the mire. Daniel pulls me out with a *squelch* and much slithering, just as the digger gets a foothold on the trunk and lurches forward onto firmer ground. At that very moment the gate squeaks. We look up and the patriarch appears. No one knows how long he has been watching our mud-larking – if indeed he had taken in any of our antics with the six-ton psychopath, which now immediately stalls.

Next day, Alan phones Digger Man. 'It's maybe bigger than we really need ...', I hear him say.

It turns out that Digger Man does have something more suitable – more delicate, lighter, better-tempered, younger, less crabby. It is a three-ton Kubota with caterpillar tracks and a permanently wet seat due to a missing windscreen. It trundles compliantly down the inclined plane from the metre-high transporter and sits smugly in our yard. Money changes hands – Alan is cagey about exactly how much.

Now the big old Hitachi must be loaded, the home team withdraws to a safe distance and leaves this to the new owner. The engine starts with only mild tinkering and the machine edges onto the metal ramp with alarming creaks and bowing. A little further up she starts to slide backwards with a grating sound. Digger man accelerates; digger slides more rapidly, grinding and squealing, and now she is not entirely straight. There is a loud clang as the spikes of the bucket come down onto the flat-bed of the lorry, where it grabs hold and the sliding stops. Digger then hauls herself up the ramp, puffing out blue smoke.

Safely manacled with massive chains, and the driver congratulated, killer digger is driven away. Alan, who is sad to see it go, watches until she is out of sight – his big, powerful, irascible friend. Now he is getting to know his new little helper, which is spinning like a ballet dancer about its own vertical axis with its arm flexed elegantly above its cab.

Our power comes from the National Grid, in a roundabout way, through overhead wires looping from rickety pole to rickety pole, all standing at a different angle to the vertical with no respect for the landscape. They are in constant conflict with the trees. The larches like to loll on the wires, and the oaks, when they have a bit of die-back, will suddenly drop a surprisingly heavy branch onto the wire where it will grab hold, sway in the wind and bring down an adjacent pole or two with a crash and a flash and another black-out. We are at the end of the line – we get a lot of power cuts.

Crisp winter mornings with a heavy frost but brilliant sunshine are best for generating power from our new photovoltaic cells. Got to dash, got to look at the meter – it has a little house icon and today it has a smiley-face. That means the water's hot and we are exporting power – our newly-minted electricity is flowing into the National Grid: that's bad – we should be using it – it's free – got to put the washing machine on – WE'RE GENERATING!

Whoops, no we're not – the sun must have gone in. I'll just pop out and look at the sky. We might manage a short wash later, between clouds. Oh no! Look! We're importing! Oh Alan, the smiley-face has gone. Something must be on! I'll check that all the lights are off. Is the fridge motor running? You haven't put the kettle on, HAVE YOU? The computer's on? Oh yes, so it is. That's it – I've got it – acute green-energy dementia – solar psychosis! I think that's how they work, the solar panels – not so much by generating, as by focusing you on switching things off.

~

People often ask us how we cope so far from the centre of things – hospitals, Greggs, John Lewis? What about the emergency services?

Have you ever tried sliding *up* a slippery pole – it's not easy, and that's why you can't un-call the fire brigade. Once they set off, bells ringing and sirens wailing, they are totally committed, and it would be churlish to stop them.

Yesterday we had a grass fire. Unbelievable after six months of incessant rain, but I've always said that Mid Wales is well-drained, and we've had a chirpy breeze in the last few sunny days. We've even generated a little electricity. We've been out and about, trimming back the hedges so they don't poke you in the eye during lambing, and sweeping up the last of the raked-out moss. A good time for a bonfire.

One little splutter from the heart of the fire: that's all it takes. Perhaps a superheated egg that slipped in with the straw from the chicken coop, or an ink cartridge that tumbled from the not-too-tidy desk into the waste-paper basket with all the bank details that must be burned. Or a tiny little aerosol can that lurked in the bag of rubbish that was brought up from Cornwall because they had missed bin day, as you do when you are young! Anyway, there is a bang and something small and very hot flies from the fire onto the bank.

The next thing we know, there is a pool of low-level flame engulfing my stamping husband.

'We need water!' shouts Alison, who is pregnant and has come to stay, for a rest.

We fill up a bucket then realise that the fire is near the stream, so run towards it with buckets. We make a human chain, but it only has two links and angina rapidly ensues as we run up and down the steep bank, up which the suddenly stiff wind is wafting the flames with amazing enthusiasm. It really makes one believe in the mischievous dryads that get bored living in the peaceful oak wood and puff out their cheeks to fan the flames and make the mortals run about for their amusement.

Alison's husband, who is stamping and beating the flames with a branch, is now disappearing into a pall of choking smoke, and the other link in my human chain is chasing her dog who has come to join in.

'It's out of control!' shouts Ali.

Now there's a moot point here – she could have meant that the dog was out of control. But the situation looked pretty dire to me and the temptation to have a little run on the flat was too much for my bursting chest, so I ran to the house to call the fire brigade.

'Emergency – which service do you require?'

'We've got a grass fire, out of control!' I pant.

'Do you require the police, ambulance or fire service?'

'Why would I want the p …? Oh yes. Fire service!' (you can tell I've been trained to deal with crises).

Now, I have not consulted before taking this action. I am usually a team player and I admit that this is not a simple oversight. I know that my husband will always argue against involving a third party – even as he is being transported from me on a cloud of smoke, he would be saying, 'Nonsense! It'll be fine.'

I have taken a unilateral decision for which I will be chastised for the rest of time … Especially as, when I return to the scene, the men have equipped themselves with a spade and the large yard broom, and at last appear to have the advancing edge of flame under control – although my broom is smoking.

I run back to the house and that is when I discover that you can't un-call the fire service.

All I can do is put the kettle on.

Seriously though, our fire service is voluntary. They come

very quickly, and we are very grateful and sorry if *I* wasted their time (please note the personal pronoun).

It does not seem to have done too much damage, but has revealed something very interesting.

The wild fire was in our barn owl habitat. It should be tussocks of grass growing through a thatch of the previous years' hay, providing cover for voles which are more-or-less the only thing that barn owls like to eat.

After a dry week this loose weave of hay had dried out surprisingly, and the fire spread rapidly.

From the other side of our valley, you can see where the flying ember ignited the hay on the side of the track. Then it spread in minutes across the 50 yards or so of rough grassland, up the hill.

Where the weave is trodden in the animal runs, trampled by badgers, foxes, rabbits, squirrels, hares, domestic cats and dogs and the occasional stray sheep, the drying and, in consequence, the burning is less. Now you can clearly see evidence of the frenetic activity, mainly nocturnal, that shapes this landscape. The hillside is black and charred but all the tracks are clearly marked on the view as hay-coloured lines that criss-cross the hill.

The fire has taken the lid off the vole habitat.

It has exposed the labyrinth of passageways, burrows, tunnels and storerooms beneath and within the sward – vole-sized ones and tiny shrew-sized scamper-ways, occasionally enlarged by pursuing weasels or torn open by buzzards.

I have found caches of lightly roasted hazelnuts, larger ones presumably hidden by squirrels but fortunately no bodies – it seems the fire moved quickly and superficially and, I guess (well, I hope), the residents fled to their basements.

We really value our dark sky, not during the day, but at night when the milky way straddles our valley and zillions of usually invisible stars make a confluence of light-points over our heads.

Getting around at this time of the year is never easy.

My problem is that when I venture out alone it is usually dark so I can't see the little icons on the second, magic, gear knob, the one that engages the four-wheel-drive. So, after I've had a little slide, I have to stop and look for the light switch, and maybe also my reading glasses. But the problem is, for the occasional night driver, when you get your head in just the right position to see the hieroglyphics on the knob, your own shadow falls exactly on that very same knob – uncanny! You can drive 100 miles at night in Wales and only see five other cars, but when you are stopped on the back road to Stay-a-little, rummaging for your reading glasses, another car will blind you with its headlights and, finding you stationary on a mountain pass in the middle of the night, the driver will get out, or at least wind down his window, to ask you if you are alright. He discerns with his knife-sharp perception that you are definitely *not* alright – but then, probably you never were.

I go to Delife to have my Welsh lesson. It is beyond Stay-a-little, a very friendly little community, albeit a tad exposed and chilly in winter, huddled in a fold on the west side of our hill. Here the weather comes in from the Atlantic, whistling up the Celtic Sea, carefully avoiding Ireland, to dump its full ferocity on the Cambrian Mountains where we live.

You wouldn't think that language had anything to do with altitude, but it has. In the sheltered valleys of Mid Wales, it is mainly the road signs that are still bilingual. The indigenous population was long ago polluted by generations of English-speaking incomers, who passed by on their way to Ireland or came to build a reservoir, but dallied, drawn by the beauty of the place and the passion of the people and the strangeness ... Nothing is more sexually enticing than strangeness (good old genetics – it just loves difference) so they fell in love and stayed, intermarried and, with the collusion of the British Government, brought up their children to speak English.

Farmers, on the other hand, are not so aroused by difference, by genetic variation. Nothing is more alluring to a farmer than the prospect of 300 acres of prime pasture, and so the farming families who are rooted in this land have not intermarried to the same extent. They live on the hills where the sun shines for more of the day than it does in the valleys, and they still, by and large, speak Welsh.

The frontier between these two foreign lands, with their amazingly different languages, runs around the edges of the hills at about 200 metres, and that is why I go up to Delife for my Welsh lessons. Although the Government pays lip service to the promotion of the Welsh language, with the recent round of cuts, my previous class folded. But, up in the hills where neighbours still chat in Welsh and the sort of folk who settle there want to join in, a kindly lady minister is running a class in a pub, without training (I assume), or vetting, or funding, or overheads, or fees, or forms, or appraisals, or even cake – it is the only class that I have ever come across that is *not* struggling for numbers.

Dyna beth od – Tybed pam! (That's odd – I wonder why!)

Square sheep

This year, Alan and I have a narrow escape from the horrors of the weather that wreaked such havoc on our neighbours' farms. This is for no other reason than that we are mindful of our own inadequacies and of the fact that when a lambing becomes difficult, we may well have to call a friend. Or even, heaven forbid, the vet. We try to lamb later, when everyone else is just about finished and is less busy with their own problems.

Eighteen sheep jostle me as I try to count them again. 'Stand still! Sixteen, seventeen, eighteen.' There should be 19 – 19 beautiful (well to me they are) pregnant ewes of the hardy Welsh Mountain variety. 'One of you is missing – what's happened? What's going on?' No one answers. Well, they all do, they *baa* but they are looking curiously at me.

What's the hold up! Get the nuts out! – that is what they are *baaing*.

Sheep always stick together. I scan the hillside. On the crest of the hill, only just visible, by the edge of the field where the oak trees overhang, there is something ominously wool-coloured.

'Oh no! I can't bear it!' Only the day before, I had boasted about our low mortality. I had tempted fate. Pride comes before a fall. Axioms jostle truisms in my head as I stride up the hill pursued by *baas*.

By the time I reach the gate of the top field I can see the large, motionless body of a sheep, with its legs in the air, like an upturned coffee-table. 'Please God, not a dead sheep.' At that moment one leg gives a twitch – I run the last 100 yards up the hill. Is she sick? In her death throes? or is she *cast*?

Cast is when a sheep gets onto her back; for some reason sheep don't work when they are upside down. It's quite an advantage for the shepherd – if you want to do something to one, you can turn it over and it won't struggle. It's not such an advantage to the sheep. Once they get onto their backs, they can just lie there giving a few little kicks until they die.

This sheep isn't dead. She is hugely pregnant. I check her ear tag: 9229. She is Square Sheep: that is her name. They are not supposed to have names, but she is one of our oldest and cleverest (though not today), and she has a magnificently heavy fleece which makes her look almost as wide as she is long: hence Square Sheep.

I gently and slowly roll her downhill until she is the right way up, and she struggles to her feet, staggers sideways, falls over and rolls onto her back again, straight back to inverted coffee-table.

This time I roll her to *nearly* the right way up and hold her there for a few minutes, talking to her encouragingly and thinking about twin-lamb-disease and the 'staggers' and all the other falling-over conditions that can afflict a sheep. Once she has calmed down, I loosen my grip and move away. She struggles slowly to her feet and stands for a while before moving away unsteadily, tacking and with splayed legs, like a sailor back on land after a long voyage.

In the distance a quad-bike revs – the cavalry is coming. And below, the other sheep stand, an ovine smear across the field, watching us walk slowly down. All eyes are fixed on the old ewe as her confidence increases and her dignity returns.

Next day we have to leave our lambing flock for a few hours: it is an imperative (life is full of imperatives). A friend has agreed to come as a locum (in between lambing the last

180-ewe batch of his own sheep), but he won't get here until after we have left.

I rise at 5.15am to check and feed the flock: chaos reigns.

Two ewes are fighting ferociously over a newborn lamb that is trying to suckle from the younger one, Number Nineteen. Every time the lamb gets near the teat, the older ewe, Square Sheep, interposes herself with frantic *baaing* and butting of the younger ewe. I chase her off, but she will not leave the lamb. And with her four-wheel drive and superior power-to-weight ratio, I am not going to prevail. I look around for inspiration.

All I see is a square, woolly bottom. A long silken thread of liquor glistens from it in the morning sun. Square Sheep has given birth! She is right – it *is* her lamb. She looks at me accusingly, and who could blame her? Still, the battle rages.

The fence is nearby. I run down the steep hill to the barn, 200 metres away, and return with a hurdle – a galvanized fence panel, two metres long and quite heavy. Then I get another one and a pocket full of baler twine. I tie them in a 'V' to make the apex of a triangular pen with the fence as its base.

At this point there is a brief intermission in hostilities. Square Sheep lies down suddenly and heaves out a second lamb which Number Nineteen licks. Then she looks at me, making the purring call that sheep make after birth: *Look, I've got another lamb – you see, they are mine!* Square Sheep struggles to her feet. This is her tenth lamb – she doesn't need this hassle.

Hostilities resume. Lambs are knocked in all directions but now I know what to do. I grab both lambs and bundle them into the pen. Both ewes stop and look at me as if to say, *That's a good idea, now let me in.* I open the apex of the triangular pen to let in Square Sheep, but Nineteen hurls herself into the pen. I secure it with us all inside and stir it until Square Sheep and the two lambs are on the far side, then I open it and give Nineteen a monumental shove and eject her.

Nineteen now dances around the pen, distraught, wailing and I have a sudden nagging doubt: it could just be that the first lamb *was* hers. I'll have to examine her to see if she has just given birth.

We have a permanent pen by the house, but how on earth am I to get her there?

I climb out of the pen and lean over and pick up the first lamb, let Nineteen sniff it, and start down to the house carrying the lamb and encouraging Nineteen to follow. A third sheep now starts to wail further up the hill and Alan comes out – it is time to leave.

With lots of running back and forth and sniffing and bleating and *baaing*, we get down to the other pen and get her in. I run up the hill and return the lamb to Square Sheep, pending further tests, then run down. The other (third) sheep is now wailing more urgently; my husband is tapping his watch. I press Nineteen in the pen, inspect her pristine, dry and tightly-closed vagina and boot her into the next field, metaphorically.

As I run up the field with a bucket of water and some feed for Square Sheep, by way of apology, I notice the wails of the third ewe are now closer together and more imperative.

Now I apply myself to the wailing ewe. She has been lying on her side in strong labour but has now rolled almost onto her back with her legs kicking in the air, which is a bit of luck because I can catch her more easily. I fall upon her and turn her on her side. She tries to get away but there will be no second chances – I am not letting go. We roll over as she pulls me down the hill, but she remains in my tight embrace. We lie panting when the cavalry arrives to hold her head end.

The lamb is well-positioned, just huge. I free its head with the next contraction, which shakes liquor all over my face and the half-born lamb *baas*. It needed a big pull to deliver the body, which is presented hastily to its mother who licks it.

We rush off to our hospital appointment, face and hair still splattered with the magic liquid.

Around the time of delivery, it is the smell and the taste of the liquor that switches on the maternal behaviour in sheep, and probably in humans. That is how a curious young ewe, like Nineteen, can accidentally get bonded to the wrong lamb. She is nearly due herself and programmed to sniff out her own lambs which might be born in the black of night.

This love potion is powerful stuff.

What's happened to poor Nineteen? She's fine, and, within 24 hours, she has twins of her own.

Magic and the lamb with two tails

Torrential rain all night. Sodden ground but not all that cold. Not good lambing weather but 'not a dying-day', Old Glyn says as we both look out across the valley and, for a fleeting moment, the sun comes out. That is what we need to hear.

Later in the day the rain is still holding off, and someone is *baaing* loudly up by the hedge. Like humans, some of our ewes labour stoically in silence, perhaps with the occasional muted grunt at the very end; but some labour vociferously. Number Twelve is a pretty, young ewe, lively and highly-strung – she shouts in labour. Today she shouts that she is at the end of the first stage, and I run out with my binoculars to supervise (that is our arrangement).

They lamb out of doors but not in the *laissez-faire*, survival-of-the-fittest way. We watch and only intervene if they need it, and if they *do* need it there seems no problem in them accepting it. I guess it's all in the timing, but our days of chasing the two-headed sheep are hopefully over (that's a sheep with its own head one end and its lamb's head sticking out the other).

Ovine obstetrics make me think of childbirth before the days of modern medicine, when more deaths were caused by officious intervention (with dirty hands) than from the complications of birth. We watch, and the more we watch the better we grasp what is normal and what is unusual for our individual animals, and we do it quietly and from a distance. Just like humans, a relaxed and confident mother is the key to a happy outcome. Also, this is a diversion for us, and we have plenty of time.

Next comes our friendliest ewe. I don't know why she is so tame. She's never been singled out for special treatment, not bottle-fed and never ill. She takes herself off into the hedge (as they do) and silently produces a male lamb.

Friendly Sheep has an immense fleece (descendant of Square Sheep) and has thick wool all over her udders. She is perfectly adapted for life in a testing climate, but her hirsutism presents a problem for her lamb: lambs are drawn to the teat by its smell and its heat – insulated teats are hard to find.

In the midst of this hunt, while I am considering how to wax a sheep's udder (ouch!) something else occurs. Something falls to the ground and rolls down the hill – it is a second lamb, and the mother is completely unaware of it. When it bleats, she looks up for a moment then goes back to nosing her first. Second Lamb shakes his nose free from the membranes with an extravagant gesture and bleats again. Again, no response.

I go and pick up the lamb and clean its face with my hand, then give it to the mother who looks pleasantly surprised and interested, and she starts to lick it while I grab Number One Lamb and we go hunting the teat. I plug it in and beat a retreat.

By 10 o'clock at night, Second Lamb is teetering about the hillside, metres from its mum, bleating weakly. I take it to its mother. *Not mine*, she baas and gives it a gentle butt, then a not-so-gentle butt.

I try again. *Not mine – smell it!*

I do. It smells terrible. Like something a dog might roll in (which is what it must have done on another roll down the hill).

When my husband gets home from his domino match, dropped off by a farmer friend, he says to the friend, 'Oh God, you know what'll happen next, it'll be in our wet-room.'

'It is already!'

I tell him that I have prepared a pen in the shed, and the friend offers us an 'adopter' – a sort of anti-butting crate. He will drop one over in the morning. We leave the second lamb in the warm, and the next hour is spent slippy-sliding up and down the sodden hillside in the rain with Number One Lamb bleating in a bucket and Friendly Sheep following me down, then panicking and running back up the field for something she can't quite remember.

The other ewes are *baaing* their conflicting advice. Eventually our old cade lamb, Aby, comes to the rescue and walks with us. Friendly Sheep is reassured and follows to the pen in the barn where we reunite her with Second Lamb, having warmed it and given it a bottle of colostrum then washed it under the tap, dried it with hay and rubbed it on its big brother so that it smells more-or-less of lamb.

Friendly Sheep settles immediately in the security of the shed. She knows she has two lambs who both now smell right, and she is letting the now-vigorous second lamb suckle.

Next morning they are a picture of domestic harmony. The adopter is not required but we will keep it until we have finished lambing.

Today, I'm late feeding the ewes and they all come running down hungrily to meet me, even the one who gives no hint that she is labouring until the very last minute. I'm not sure she knows it herself, but something stops her about 50 yards away from where I was busy with the cratch. As I fill the trough with concentrate, I am aware of her in the corner of my eye.

She halts, looks surprised, and with one huge effort pushes out a large lamb. I can see the lamb moving but, moments later when I get there, things are not looking good. I pull a great wadge of membranes from his throat, but he does not react. I swing him then press rhythmically on his chest. I blow into his nose and poke a grass stalk up his nose. I even do something that I really shouldn't do: I give him mouth-to-nose ventilation like I used to do with human babies, born by caesarean section and reluctant to start breathing for themselves. That was before we worried about infection. It always worked with them, thank God, but it doesn't work today – he is resolutely dead.

What a waste! What a blow! Fifteen seconds earlier, half a minute perhaps, and it might all have been different. Although maybe the precipitate delivery of so large a head caused a haemorrhage in his brain. Anyway, if there was a window of opportunity, it is now closed.

I start to take the dead lamb away and the ewe wails, so I put it down again. She stops wailing and I think: I do wish she could have a live lamb. Then I remember!

I carry the lamb into a pen and the mother follows. She looks worried and paws him a little but is not so distressed – perhaps she thinks he's having a sleep? She sniffs at the bucket of feed and I go to make a telephone call.

A few hours later her new lamb is suckling. In the half-light of the barn it is difficult to see that anything is out of the ordinary – he is wearing a pink sheepskin overcoat!

Thanks to two neighbours, this little orphan lamb now has a healthy young mum with lots of milk. And the ewe – well, she is none the wiser. Her lamb that would not move, nor *baa*, nor feed, that lay in the pen unresponsive to her pawing, was taken away for a moment by the big man who comes in the red truck, and the next minute her lamb is right as rain – so right, he has two tails!

A little deception and the application of an old country skill, and the dead lamb was discreetly skinned and the skin with all its odours and associated love is transferred to the orphan (acquired from the other neighbour). And everyone is happy. The skin fits like a jumper. And, as a finishing touch, the placenta and membranes, cleverly collected in a clean bucket, are rubbed over the new lamb which is why it is pink.

This practice is common and makes all sorts of economic sense. The lambs do better and are much less trouble. Formula is expensive, and the bereaved ewes don't waste their milk or get mastitis. But that's not the reason that trucks and gators dash about the countryside at the busiest time of the year carrying bleating boxes.

The reason is empathy, sympathy. Dare I say it – love. The toughest heart is softened by that wail and the hungry lamb's bleat. Even the most practical farmer has to acknowledge the emotional bond between mother and baby. It also affirms that Biblical relationship between the shepherd and his flock.

The following day it is impossible to remember the sad little orphan texel-cross lamb who came to be adopted.

He put on the mantle of a much-loved but non-functional Welsh lamb and confirmed our friend Gareth's reputation (at least with one ewe) as a miracle-worker.

Three days later the lamb's magic overcoat is removed, and off he bounds (considerably less smelly) to join his peers, to the obvious satisfaction of his proud mother.

~

Two new lambs are born before dawn, they are lying with their new mother under the hedge – both healthy. Above the hedge I spot another ewe. So, before marking the new lambs, I go to check the other ewe. She is licking the ground and chewing on membranes in the grass. From her rear dangle other membranes and her large udder is streaked with blood. She has obviously just given birth, but there are no lambs.

I hunt up and down behind the hedge – there is no trace. Several neighbours have been troubled by a predator this year.

Damn! DAMN! I should have been up earlier. That wretched fox has had a newborn lamb ...

Unless ... something in the manner of the ewe with the two lambs, below the hedge, is not quite right. As I approach her, she looks excited, not wary. She gives me that *Oh-good-it's-time-for-breakfast* look. A sheep that has just delivered usually has more pressing things on her mind.

I rapidly fashion a pen out of hurdles (having left them in the field) and lift the new lambs into it, then let the new mother in and examine her rear: it is clean and dry – she has stolen these lambs. More accurately, she has kindly fostered them after they rolled through the hedge, probably because of over-enthusiastic cleaning by their old mum. Energetic licking is prone to start avalanches on our steep fields.

I return the lambs to their birth-mother who looks doubtful. She smells the first lamb and nuzzles it, but pushes the other gently away. It rolls through the hole in the hedge and bleats. Foster mother screams from the pen and tries to jump out, collapsing the whole caboodle. The lamb rushes to her and suckles.

I sigh and try a different tack. I carry the rejected lamb down the not-inconsiderable hill (up and down which I have now been running for some time). The foster-ewe follows me, complaining, and I shut her in a more substantial pen. Then I repatriate the lamb, which is surprisingly vigorous, with its real mother. *Not mine!* says the real mother and knocks it over.

'Yes, it is.'

No, it isn't – look! She's pushed it through the hedge, and it is running amok, bleating, and several of the other ewes are coming up to investigate. *Not mine!* she insists.

That's it! I've had enough. I bundle the troublesome lamb over the fence, reunite it with its real mother and sister and then we painstakingly walk them, with much arguing and to-ing and fro-ing, the long way round to the barn, where I shut up mother and both lambs in a small pen.

After such a long and tiresome walk so soon after giving birth on the frosty hillside where it is now raining again, the sight of a warm, dry pen and a bucket of feed persuades the mother to concede: *Alright they both* might *be mine, but I still don't like the look of that big one.*

She has now fed both lambs and Alan has brought me a cup of coffee, but still the cries of injustice from the kind, obliging foster-mum can be heard – I hope she will have her own lambs soon.

And to reassure any farmers reading this, just to be absolutely sure, I go and find the placentas and they are both *above* the hedge. I don't know where the midwife has gone – she's the magpie who usually clears these things up.

The following morning things are not looking good. The mother is upset and butting the larger lamb who is starting to look wary of her.

Now I am getting wise to this bonding business and I pick up the big lamb and sniff it – she smells terrible, like rotting fox poo or something dead. I sniff the little one – she smells all lamby and nice. Cracked it!

So, while Alan fixes the lamb-adopter to the pen, I wash the offensive creature with clean, warm water. She doesn't like it much. Then I dry her with odourless kitchen paper and finish her off with her sister's woolly back. Then we exercise the human lamb-adopter who has come to investigate. He holds the ewe and the big hungry lamb has a feed and we squirt her with her mother's milk. The mother is sniffing them both now and looking confused – hopefully she can't count.

The lamb adopter is a hurdle with a V-shaped cut-out in the top and a retaining bar that fixes across, converting the 'V' into an inverted 'A'. All you have to do is fix it firmly in the pen and persuade the ewe to rest her head in the V, then bring down the bar and trap her. This is best done by cunning, rather than by force, and you should clench your teeth while doing it, so that when a clever ewe butts you, you do not bite your tongue. Place buckets of food and lots of water on the head side and the lambs on the udder side and there you have it. She can't butt the lambs and they can feed from her freely.

After two harrowing days (for me) in the adopter, the ewe can't tell her large, washed and dried lamb from the smaller lamby lamb and they all leave the shed together, like a proper family.

Interestingly, the mother seemed less stressed in the adopter than she had before when she was having to constantly fight off smelly, alien lamb to protect her lovely lamby lamb from the atrocity.

Also, she liked the catering arrangements in the adopter – so much so that at feeding time she now brings her two lambs back into the shed and into her erstwhile pen to be fed, and so that I can weigh them (sheep like routine).

The rest of the time they are outside with the other twins.

Welsh hill farmers are not generally known to launder their lambs. I have an excuse for this aberrant behaviour: I remember the glorious smell of my own new-born babies. I always suspected that one reason why some human mothers failed to fall in love immediately with their babies was to do with the burning desire, of the midwives of my youth, to wash the babies in their care. If a new mother was conscious, her arms were the most convenient place to park a newborn baby while they cleaned up the mess and filled in the forms. But if the mother was ill or still asleep after having an anaesthetic, no-one wanted to leave a baby all streaked with blood or worse ... 'We'll have her all spruced up and smelling of talcum powder before Mum wakes up!' It was well known that instrumental deliveries and C-sections were a risk factor for failure to bond. At that time nobody had worked out all the reasons why.

It's fundamental, but then sometimes the obvious is overlooked or forgotten.

Domestic deity or just a damned cat?

'Don't get up and feed him the instant he meows!'

'But he's hungry.'

'He's just a damned cat!'

Alan is not a cat-person. Here's the dilemma – the main cause of tension in our household.

Guinness, The Fat Cat, epitomises the power of self-confidence. He strutted into our house three years ago, stood his ground when the dog rushed up to him. Their noses touched for an instant, the dog was transfixed, then *Wham*! With the paw of steel, the dog was dismissed, blooded: dominated.

Guinness moved in with his own household – his man, actually a cat, Midnight.

Cat psychologists say domestic cats are solitary. That is nonsense. Guinness has a butler, his own Jeeves: someone to see to his personal grooming, to suffer fur balls on his behalf, to hunt for him, to taste his food, to intercede with the other servants (me and the dog) and to do his meowing.

Midnight – 'now he's a *proper* cat!' says Alan – is The Fat Cat's batman. They met in a previous life, not in the army but in prison – a prisoner-of-war camp, *Stalag 46*, in Brighton, in the war on the Feral Feline Freedom Forces. The Fat Cat was in charge of escape plans, and very good at it. So confident was he that he would present himself every morning *outside* the prison, at the camp commandant's bungalow for breakfast. After breakfast he was marched back to the pound where the other prisoners greeted him as a hero, the Steve McQueen of the Cat Rescue. He would eat again and sleep all day, with Midnight, The Proper Cat, watching his back.

The Fat Cat and his side-kick were released on licence to live under house arrest in Worthing, that pit of iniquity. (I was chased by a mugger once in that East Sussex town fallen from glory. Once respectable, it became for a time a forest of parking meters roamed by drug addicts, prostitutes and cats – or so it seemed to me.)

Still uncontainable, it was here that he forged links with the underworld, colluding with local foxes, pimps and minicab drivers. Wandering the streets at night, his were the green eyes under every illegally parked car; his DNA was on every discarded take-away carton.

He came to Wales under cover (he's a sleeper – don't tell anyone). An urban gangster lying low. Some say he worked for a Russian bank (no one knows the full story). Now, he's free to come and go, he mainly does what he does best (he's a sleeper after all). Under his protection, Midnight, his faithful lieutenant and a 'proper' cat, does the rest. They are Farm Cats Ltd. (Non-exec. Chairman: Guinness, 'The Fat Cat'.)

The tax-man knows we have two farm cats that control the vermin. Midnight does that. He is sleek and black. He catches the mice, voles, rats and the odd mole, while the other, Guinness, does the managing. He is an agent if you like. He takes a cut of the quarry, and a percentage of the pay (60%, I think). He's never gone out much, but he coordinates from his office by the wood-burner, while Midnight is out in all weathers ... That is until recently.

A little while ago I came down in the morning and stepped over Guinness, sprawled in front of the fire, basking in the heat. But hang on a minute: the fire is out.

'Is that cat dead?' said Alan. And I'm afraid he was. It was all rather unnerving and sudden. Though he had climbed a 15-foot pollarded tree the previous weekend, which was so out of character that Alan had commented that he might have a bucket-list.

Perhaps he did.

The amazing thing is the change in Midnight, the worker. He didn't go out for three weeks.

'He must be grieving.'

'No, he's not. He's inherited the territory, the house, the staff, you and me.' Always a cat of very few words, within weeks he is waking us up, caterwauling at the bedroom door, demanding food, chatting, complaining about the weather, knocking my handbag off the kitchen table if I put it where he now likes to sit – I don't know what will happen when the spring comes, and all the vermin start to reappear. Perhaps he'll advertise for an assistant.

But Midnight has been helping me with my research.

You see, some of my best friends look very different from me. I spend a lot of my time with individuals who see the world through very different eyes, and I would like to understand this better.

For a start, their eyes shine at night, not with avarice or the holy spirit but with any light that they catch in their eye.

You see they, the sheep at least, are a prey species, and they stand out all night in the darkest fields, uneasily looking out for wolves and rustlers, so they need to see in the dark. One of the adaptations that many nocturnal mammals have made is to acquire a *tapetum lucidum*, a biological mirror behind their translucent retina, so that light stimulates the retina as it falls upon it, and stimulates the retina's photosensitive cells again as it bounces off the mirror-layer, heading back the way it came. This helps them to see in the dark.

So, when you go out in the field at night with your torch and all the sheep turn to look at you because they think you are something spooky, all their eyes light up with intense pale green light, all directed at you, which is quite impressive, even sinister.

Dogs have a *tapetum lucidum* too – our dog shines bright green.

Foxes have eyes that glow green: different species have variation in their *tapetum lucidum* and glow differently. Hunters who go out lamping for rabbits and foxes (I think it's illegal so they can't be doing it anymore) tell me they can tell what they are shooting by the shade of the eyeshine, as they charge around in a truck with a lamp on top, picking up eyeshine and shooting things (fortunately humans do not have a *tapetum lucidum*, otherwise more of them would get shot).

Cats are famous for their glowing eyes and that is where I get into trouble. I have just spent a happy evening flashing and photographing our cats, trying to demonstrate their eyeshine, and their strange lozenge-shaped pupils that constrict down to a tight vertical slit in bright light (you see, one of the problems for these creatures, who are adapted for the dark, is managing bright light). Mainly they shut their eyes. This makes them seem very relaxed and contented, which endears them to their human carers who think they are smiling.

It is only a few minutes since we finished my unsuccessful photo-shoot. Midnight (our short-haired black cat) has started doing something very strange and alarming, kicking his right foot out then grabbing at his mouth with both his front paws as if trying to pull something out of his mouth. He does this repeatedly, making a peculiar squelchy noise. There isn't anything in his mouth or throat. He isn't salivating or retching and there is no sign of a bite or sting on his lips. The other cat and the dog look worried, and are following him around, fussing as he repeats his odd, stereotyped gestures – like a non-verbal Tourette's Syndrome. OMG! – he's been out and got a head injury, or a brain tumour ... or epilepsy due to flashing lights. Now ensues a period of research on the internet. While the cat twitches, quietly now, on its chair by the fire, the other two animals sit upright on the floor next to him, watching anxiously.

By the time my husband has got home, I have cracked it – Feline Hyperaesthsia Syndrome. Apparently, it 'can be provoked by stress' (like being chased around the house with a flashlight). This is a diagnosis of exclusion, and mindful of vet's bills we adopt an expectant policy: we'll watch and expect it will get better.

It does. For 12 hours or so he looks spaced-out between twitches that gradually get less pronounced and with longer gaps between them. First the kicking disappears, then the grabbing at his mouth, then the licking of his lips gradually stops, and he has a long sleep. He wakes up and has a large breakfast and is now fine again. We didn't photograph any of this (I think I have done enough harm).

Returning to the great mysteries of the mammalian eyes that follow me daily, I ask why do cats have vertical pupils and sheep horizontal ones?

They both need to be able to restrict the bright light of the midday sun. Cats need very sharp vision right in front of them, and the potential to use a whole cross-section of their lens (this has complicated optical reasons to do with putting back together the spectrum that bending light tends to produce). Thus they need a vertical slit because they are predators, and they pounce on little creatures right in front of them.

Sheep need a more global view of the world – they live on grassland and need to be able to spot movement all around. With their protruding eyes and wide pupils, sheep can see from right in front, to right behind, back along their flanks. Provided they walk in a slight zigzag (which they do) they can see all around themselves, even in bright weather when their pupils are constricted. They couldn't do this with a vertical or a round pupil. Presumably we can cope with a round pupil because we have heads that swivel about a vertical axis!

Dogs have round pupils. They are generalists like us, but can see in the dark. They have reflectors at the back of their eyes, which shine but are not so sensitive to the light that they need slit pupils to protect themselves by day (I suspect this is because at night they see mainly with their noses).

If you watch a blind dog with four legs and with other smelly friends running with him, he can charge about the 'odourscape' with alacrity, avoiding the objects which must cast a scent shadow, or smelly aurora, in their aroma-mapped world. It's amazing!

Think about bats: they map out the world in their tiny heads, not with light as we do, or smells as Pedro might, but with bouncing sound-waves. The processing is similar, just the sensory input has changed. The mole: he does it by touch. And the catfish, in black water, pictures his world in taste and feeling. It is almost incredible.

Pink balloons

Sheep can undoubtedly tell the time. However, I fear their grasp of numeracy is in doubt. Number 39 is a good mother – she raises one fine lamb every year.

This morning, she proudly brings this year's lamb down to the gate for me to admire, and to let her through to the other field – the post-natal ward, where generous extra rations are available to aid lactation.

'But wait, 39, whose is that?' I point up the field.

Just born and all alone (too young to be all alone) wandering about under the hedge, 50 yards away.

This year she has had twins, which confuses her. She knows they are both hers when they bleat or come close enough to smell. The trouble is she can't count. When one goes to sleep, she wanders off and forgets it.

She looks at me accusingly: *What do you mean, where's my other lamb?*

Thus, amid a ferocious hailstorm, I am seen running across the field with a wriggling lamb under each arm, hotly pursued by an angry ewe trying to knock me over sideways. Anyway, the penny drops that I am not trying to abduct them when I plonk them both in a nice, dry pen where mum is happy to join them and start her crash-course in remedial numeracy – we're only going up to two this year and she's picking it up already!

Our hardy Welsh Mountain Sheep aren't really made for triplets – in ten years we've only had three sets. The first set were all born dead and the mother sadly also succumbed – our biggest ever lambing disaster.

We aren't technological, we don't scan. With fewer than 30 lambing ewes, it would be difficult to arrange as we don't have the economy of scale. It's not worth it to come to us and set up the kit, and no-one wants to be bothered to come and do it. We know our stock, and if they are hungry and losing condition, we just feed them more. It's quite exciting seeing what we get – like Christmas.

Our lambing husbandry lurches between that of a pre-war hill farmer and the maternity care lavished upon humans in North London in 1976, when I learned most of what I know about obstetrics (I was a Member of the Royal College).

I remember learning that, when the mortality rate for human newborn babies was last at the level of lambs born in our fields, and that of their mothers was much higher than for our ewes, the cause of maternal death was usually sepsis due to dirty hands rummaging in the genital tract. Often there was little necessity for this, apart from the ego or ignorance of the midwife/doctor, impatience, or idle curiosity.

Thus, I try to exercise patience and self-control (which is against my nature), and an expectant policy of wait-and-see, and only-intervene-when-absolutely-necessary. Then we give antibiotics to the mother if there has been the slightest rummage. Keeping a gloved hand sterile is challenging in an operating theatre; it is hopeless in a wet field in the dark.

Just pulling out the lamb that has conveniently presented you with its little hooves does not really count. That's just a *Little Jack Horner* moment, to make the shepherd feel *what a good boy am I.* It speeds things up a bit when the mother is tired, and stops you having to shout at her to *Push Harder!* (Why do midwives do that? Presumably because there is nothing to pull. If only human baby's heads had handles).

Even in the Welsh hills, fewer and fewer farmers lamb out in the fields. Most do it in the shelter and warmth of large sheds, with pens and good lighting and enough sheep to make it worthwhile, and to have someone there day and night. For all that, I don't think many of them have wash basins or rubber gloves yet.

We are an anachronism. I stumble around our fields with torch or hurricane lamp listening to the sounds of the night. For the soft gurgling *baa* of a ewe calling to her newborn lamb, or the hysterical rhythmic *baaing* of a frightened young ewe, lambing for the first time. By and large our sheep do not lamb in the hours of darkness (one of the advantages of not having the lights on all the time); their circadian rhythms are undisturbed, even if ours are not. Our lambs usually arrive at or just before first light.

Last night, though, was an exception. It was clear and frosty, and there was a huge, bright full-moon casting weird shadows over a black-and-white world. I was woken by a bleat at half past one in the morning and went to investigate.

Up in the corner of our steep field was one of our first-time ewes running after two lively new-born lambs. I've been worried about her as she has been large and uncomfortable and has had a large swollen udder. It is bare of wool and oedematous so that it looked translucent, like a large pink balloon.

This should have made more of an impression on me.

Anyway, I was very relieved to find that she had delivered without any trouble, and that the lambs were so lively. So I went back to bed ... mistake!

At dawn Alan counts the sheep and calls me: only one lamb at foot and something white on the ground.

Sure enough, there is a dead lamb, a large one, cleaned to a dazzling white, cold and dead on the grass. That's odd – they seemed so lively; but closer inspection reveals that the young ewe still has her two live lambs. One is tucked up behind her.

After failed attempts to resuscitate this perfect, cold, dead ram-lamb, we have a roll-call and tail inspection of the other ewes in the field. No-one has any signs of having delivered a lamb. Our dead lamb is the third of triplets – all that physiological effort gone to waste.

The two surviving lambs, born in the night, are exploring their leafy environment, and I am pondering on missed opportunities and lessons to be learned:

1. Be particularly vigilant on bright, moonlit nights when the light level is high and when all night can seem like just before dawn.
2. An udder like a pink balloon may herald triplets (quite rare in our breed).
3. Remember to wait a while after delivery, as it's the last, often smallest, lamb that slips out unnoticed or sometimes just rolls off down the hill and gets overlooked when the ewe has so much new to deal with.

A year later and wiser. Number 32 has a huge, strangely translucent, pink udder, and this time we know what it might mean. It means we must watch even more expectantly for a third lamb.

She has now produced two healthy lambs. Fortunately, learning from my previous mistakes, I sit and watch all afternoon while these first two are meticulously cleaned and properly fed. Then – Bingo! Number three arrives. Not breathing and with very soggy-sounding lungs, but nothing a traditional swing or two and some frantic chest compressions does not sort out – amazing! The swinging, holding it by its back feet, really does seem to shift the fluid – I never really believed it.

Then of course I must wait till he's been cleaned and slowly fed, and then some more – we don't intend to be caught out by our first quads!

~

My past is always insinuating itself into what I do now, making me seem eccentric.

Take the example of the mystery of the missing ears.

Our friend Tony told me about the flock of sheep fed on oilseed rape, when it first became a popular crop in Britain. (It might have been in 1976, that famously hot, dry summer.) The sheep gained weight like never before, but lost their ears! It was a mystery.

Phototoxicity is something I learned about in another life when a lady gardener showed me the livid, blistered scalds on her arms, as if she had been whipped with a red-hot flail. In fact, she had been lightly brushed by the cut, sappy ends of giant hogweed, angelica and cow parsley that had taken root and flourished amongst her parsnips that hot summer and which she had been cutting down.

Phototoxic chemicals, which occur in all these plants, can

increase the reactivity of the skin to ultra-violet and sometimes visible light. They are the opposite of sunscreen and can produce the most bizarre patterns of sunburn. You can get them onto your skin directly, like the sap, or be affected by eating them, as with the light sensitivity that can occur with certain drugs.

Bergamot oil is another phototoxic agent, giving a puzzling, blistered burn on the neck of a very smart but distressed lady who did no more than spray herself with expensive perfume on a sunny day.

I think that the sheep with the missing ears got such bad sunburn on their ears (their least woolly part) after eating or brushing through oilseed rape that they ultimately lost the tips of their ears – like two of our lambs this summer.

Two of our triplets this year seem to be affected. They are the small ones who have had less milk and, early on, were hungry and foraged more widely, nibbling in the hedgerow, and tasting all sorts of plants at an earlier age than usual, when their hair was thin and their skin sensitive. The larger of the two has gained weight but has lost his ears!

The little one has done better, now that we have worked it out, and she has had treatment and gets my sunscreen (factor 50) liberally applied to her ears and nose on sunny days.

Generations W, X, Y, Z

If I put myself in the shoes of you the reader, I would probably say (as I often do when reading other folks' writing about their own lives), 'the bit that fascinates me is the bit that is missing'.

Today our children are home, and I have been thinking about family life.

Looking forward from childhood, all we ever knew was a nuclear family: Mummy, Daddy and 2.4 children (black Labrador and/or a cat – optional). Actually, I was an only one, as was Alan. As you know, my elder brother died the day he was born.

Looking back, the nuclear family was never, ever what it seemed. Now that I'm old I can tell you – as people have told me (remember I was a family doctor) – it only ever *seemed* to work, because of hypocrisy and deceit. It was a construct – like a Facebook persona. All those role models were *having you on*.

Your parents were never who they seemed. They stayed together because they loved you, or even because they loved each other, but mainly because they saw no other option – it was never easy. Your uncle and grandfather and half the men they knew had mistresses, and their wives had unhappy love affairs (sometimes lesbian), or were just unhappy, often deeply dejected and rejected, overworked and unappreciated. Your respectably married, professional cousin had homosexual adventures. The vicar was mercilessly preyed upon by a series of lonely housewives, and the headmaster was patting the bottoms of the little girls. Young lads crossed the road to avoid the priest, and mothers might be heard excusing their sons with the mantra 'boys will be boys'. Even the smallest village had a couple of convicted paedophiles and several prostitutes, if only semi-professionals – kind-hearted women, enthusiastic amateurs. And the milkman and the postman may well have lingered longer than necessary. Even the roving greengrocer was not averse to having a go – I remember my Mum's outrage at his audacity (more to do with class than virtue; the vicar would have been rejected more graciously).

All this happened. It is what the Bible called 'original sin': it's not original at all; it's human nature and you had better believe it. Accept it and live with it.

One day (I don't remember quite when) the veil was lifted. Perhaps it was the advent of the mobile phone or itemised bills, but suddenly deceit became more complicated, and women had more options. We entered the era of family breakdown. It was very painful.

The worst thing was that it wasn't supposed to happen – not to people like us. We were all so unprepared.

However, 30 years on, I can tell you that there can be a post-nuclear family, a reconstituted family – just add water and stir: children, grandchildren, stepchildren, half-brothers, cousins, biological mums, gay partners, feckless uncles, previous lovers, ex-in-laws, anyone you like! The only rule is to look after the children, and don't forget the dogs.

'Aye-aye', I hear the reader say, 'I sense a back story here' – and well you might. But it doesn't do to dwell on the catastrophes of life. When things cannot be changed, they must be accepted. We got to where we are today by being adaptable ... so, adapt!

We – Alan and I – choose to live on the edge. Wrestling with physical adversity seems to be something we need to do. It makes us pivotal in our own existence. We built our home here, with a little help from our friends, and there is little that needs to be done that we haven't learned to do ourselves. We cut down trees that threaten our electricity wires; we mend our bridge and tend the well; we poke sticks through the crust on our cesspit and nod our heads wisely. Our barn and woodsheds are full, and we have enough lamp oil to last a whole winter without power.

I get perverse pleasure from asserting myself over an animal that weighs more than I do but doesn't understand the need for immunisation. We can gather and marshal our little flock with ease now, albeit more by cunning than physical fitness. Still, it's very satisfying.

It's only when our oxymoronic, grown-up children arrive, bringing their own worlds with them, that we start to feel marginalised on our edge. It's not that they sit in silence texting (they don't anymore); or have a hybrid stealth car (we used to enjoy bump-starting their old wrecks); or that they don't want the benefit from our experience about routes home – *it's okay, Mum, we'll probably just follow the sat-nav.*

They bring their films and music, grab the remote and find channels on our TV that we didn't even know existed, channels which have their own familiars, strange creatures that leer from the box making jokes that aren't funny (to us). But they laugh – their cultural allusions are lost on us and ours on them.

They ask for mysterious bathroom products that are not soap nor toothpaste. They want to know if the cheese is pasteurised. They are outraged that the baked beans went out of date in 2006 and the cloves in 1994, and fill the fridge with essential dietary items that are unknown to us. They ask their father not to smoke in the kitchen and are shocked by the lack of a toaster and hair conditioner and the suggestion to use washing-up liquid when they finish every drop of shampoo in the house. And they look slightly put-out when one of us absent-mindedly breaks wind.

They can't understand a plumbing system that refuses to

act as a garbage disposal unit, and which exercises its own water-rationing. They put all the lights on all the time and use unbelievable amounts of toilet tissue. They download all sorts of strange programmes onto our computer, change the browser (so we don't know where we are) and complain about the broadband speed and the poor wifi signal. They put everything in a different place and worry about our poor memory and declining powers when we run round in circles looking for things.

When they go for a 10km run or climb one of our little mountains, we provide back-up or bring up the rear, panting, feeling our age, feeling bad because we've held everyone up, made everything more complicated. We are no longer central to the exercise.

Then something occurs that bridges the gap, that slots us all back into a familiar place – they produce another generation.

Suddenly we're all on familiar ground, all singing from the same song-sheet – *The Oxford Nursery Songbook* to be precise. We're all saying the same daft things that we did 30 years ago, blowing raspberries, funny voices – Tweetie-Pie is *going south for the winter* again.

Now generation Y is at the sharp end, and we can look on and smile. Another little girl is learning to be gentle with the pussy cat and not to pull the doggy's tail. Another boy is learning not to put the fish hook through his finger.

Expeditionary forces again march behind their trusty leader without the benefit of sat-nav, and trudge home tired for tea.

There is someone new to show to the sheep and to introduce to the joy of muddy puddles. Mummies and Daddies are soon entering into the spirit with a planned Bank Holiday dip in the pond – *it's not chlorinated you know.*

Auspicious day? Yesterday I cleaned the house, which must have unsettled everyone. Today I can feel it through my feet – bare feet, next to Godly feet; feel the silky smoothness of the wooden floors, the springiness of the fluffed-up carpet pile and – bless the cats – a gift: a dead field-mouse.

Just what I wanted to step on, on my birthday. But not as much as I wanted my breakfast. A bowl of muesli, topped with fruit-salad and – go on, spoil yourself – some left-over double-cream. I'll eat it in the sunshine. So, I pop it on the table outside the kitchen door. But remembering the mouse (other members of the household are more squeamish than me), I pick it up by its tail, go out of the door and hurl the little body towards the overgrown bank (for the thrice-mewing buzzard to eat). Not a very good delivery (sad in a cricketing family): *Splash!* – luxury dead-mouse muesli!

You see I've been re-reading the Scottish play. The last time I read it was when one of my daughters was doing it for GCSE, 20 years ago. And I remember how our children then sat around the tea-table competing like the three witches for the most disgusting image to toss into their own imaginary bubbling cauldron, a delightful exercise in officially-condoned impious utterances! None could compete with the bard, who we all agreed must have sat around a table in the tavern near the Globe, with a bunch of actors doing exactly this same thing. Who knew that Shakespeare could be such fun?

It set me thinking what else I should add to my birthday breakfast.

We live in Wales, so dragon scales. And it's a classic recipe, so:

Eye of newt and toe of frog,
Wool of bat and tongue of dog (all freely available).

I can get hemlock and yew,
but liver of blaspheming ... oh.
'That's racist!' interjects the spouse. 'Just make do with mouse'.
Entrails of a cursing Celt
Would probably suffice, I felt.
Nose of Turk will no longer work,
A politician's naval, I'll convert.
For Tartar's lip, a grated betting slip,
Ear hair from a defrocked priest,
Would be in the spirit of the piece.
Gall of goat and sweat of stoat,
Blood of python foamed up in a soda syphon ...
... that should do it!

Now that our daughter is in the last weeks of her first pregnancy, I have been getting to know the stranger who will soon be in our midst. The new arrival to challenge all our preconceptions and established relationships. Presently a nudge to my proffered hand, like the mumble of a presence in another room, but soon to be much more. You see, nothing is ever what it seems. You are not what you seem, not what you were a moment ago. Time changes everything.

A first baby, at any time, changes a child into a parent and sends the parent from centre stage into the wings (maybe a relief after 35 years in a leading role), to emerge in a new role, which could be tricky but could be amusing. Even liberating.

A new baby makes little sisters into big sisters; little brothers become big brothers, mentors and protectors; siblings become aunties, next in line of responsibility, in case of disaster. And dogs. Sorry, Pedro, but dogs become a threat – you had better make the most of the next few days.

We are all looking forward, or are oblivious, to our impending rebirth. Meanwhile we wait.

A fortnight later, the photo arrives through the miracle of modern technology. The image capturing the situation as words never could, quite. My daughter, pale and dishevelled but more alive than I have ever seen her – I recognise that wild, slightly shocked look of satisfaction as she shows a small, crumpled baby boy to the camera. I think *adrenaline!* I think *now I expect she will moult.* And she does! Our experiences seem so unique, but they aren't, not even to our own species.

A Mammalist's view of language

There are two sorts of individual: those who need words to express themselves and those that do not. If you are a writer, you are likely to be one of the former, but not necessarily ... *Nonsense! Everyone needs words.*

We have a friend who is a bit different. Actually, he is quite a lot different. He has a genetic abnormality that affects his ability to use language; the language part of his brain is absent or switched off.

A dog knows some nouns: his name, *dinner*, *walkies*, *stick*; and some verbs: *fetch*, and hopefully he understands *No!* My dog understands some phrases: *feed the dog*, *feed the sheep* and *go to bed*. But he can't articulate very clearly. And, okay, his grasp of sophisticated and abstract language isn't great.

Our friend's articulation is better. He has the right equipment to make the sounds, but his grasp of language is similar. This is quite a disability, but not *that* much of a disability. He looks different but is physically robust – *as strong as an ox* – he has good balance and co-ordination, is hard-working and eager to please. He will dig or sweep or wield an axe all day. He will walk home, day or night, miles across the fields and is never out of work and rarely short of money. He also has terrific social skills, notwithstanding his appearance, and strangers' often negative responses to him for all the above reasons.

The thing is, he has a very well-developed grasp of the non-verbal. He knows exactly what is going on, who likes whom, who doesn't and who would stab you in the back – 'Bad man!' he says and he is invariably right. This is another reason folks are wary of him. I'm not sure about the workings of his sense of humour but he loves to laugh, he rejoices in laughter, is attracted to it, infected by it, bathes in good humour when it surrounds him. He gets along well with Alan.

His life is very difficult. He loves to be in a social setting, but social settings increasingly fear those who are different, and find excuses to exclude them.

It's a shame I cannot transpose his perception of the world into words for you because, if I could, I'm sure it would have emotional intelligence that would stun you and there would be no deception – no horse-feathers!

Words are very blunt instruments.

This brings me to the thought that set me off on this tack. People who haven't always lived with animals find it odd, the concept of personality in other species, and this is only because we depend so much on language. It means we miss so much of what is going on with other species or dismiss it as unimportant. But life and death are always important, and the drama is just as great whether you are a sheep, an old woman or a bank vole!

In my philosophy, God and Mother Nature are mostly the same thing – *she-who-knows-best-in-the-long-run*. However – meet Gladys!

Born yesterday lunchtime and not quite right, the second and much smaller of twins, she popped out as an afterthought and lay on the grass ignored (Nature knew). Then ensued much running about, building of pens, pressing of the ewe (who also knew), and the spifflication of an elderly gentleman who held the ewe (who knew) while the elderly lady (who also really knew) tried to milk the ewe who knew. The lamb, who was not quite right, was held to the teat and made slurpy noises but nothing came and so they rummaged in cupboards and under beds and assembled the 'milk-bar'.

Meanwhile the lamb got weaker and weaker, and visiting farmers (who knew of course) shook their heads and advised euthanasia, only that wasn't quite how they put it. 'Call her Gladys', said one, amongst the more helpful remarks.

'I know she's not quite right. She's got under-bite, and no cartilage in her ears and her back is twisted like the toy lamb Alison had when she was little, whose wire frame got bent by too much cuddling (do you think she'll unbend with time?). Some babies have funny-shaped heads, but they come right ... or they get hair so no one notices ...'

Gladys takes to the bottle like a professional – and, God bless the ewe who knew, she is amenable. When the spifflicated gent stops holding her, she stands and watches me feed the lamb, and then, bemused, she cleans up the smelly, milky mess I have made of her and takes her off with her other lamb for a rest.

Last night I slept badly, wrestling with a moral dilemma of the lamb who will not do. My head rings with advice. Farmers say, 'the first loss is the easiest'. An old boss of mine used to say, 'we must not strive officiously', when he meant, 'it's time for this poor little bugger to meet her maker'.

I wake up decisive. No more feeding – it's up to the Shepherd in the Sky. The lamb doesn't seem to be breathing. It has been a cold night. I tiptoe to take her body from the pen without the ewe knowing: a little tufty ear twitches, a small black eye opens. The crooked lamb jumps up and runs to meet me, *baaing* for breakfast (still a bit wobbly).

I go indoors to think about it while I make up her breakfast.

Death is always the same but who knows how Life will turn out – that's the trouble with euthanasia.

~

It's Election week, so you might think that being born without ears is a good thing, but Gladys the lamb can actually hear quite well. When a bird squawks or the pigeon that has moved into our expensive new barn owl box starts cooing (as well it might), Gladys pricks the tufts where her ears should be, and cocks her head towards the sound.

She is very active and, although someone I thought was my friend accused her of having knobbly knees, she appears to be growing and developing normally. We bottle-feed her for three days, by which time she cottons on to the workings of her mother's udder and dismisses us. This is very gratifying.

Now she is playing with her friends, running her mum ragged. Mum doesn't notice that she is any different now. But then mums don't – in fact she gets quite muddled about which two lambs belong with her.

Sheep express themselves with the subtle waggles of their

ears and their angles of elevation, so it is no wonder that Gladys is becoming very loud. She is compensating vocally for her lack of ability in the semaphore department – the missing link to talking sheep, perhaps?

Is Gladys just premature?

Born three days ago and left for dead – a bag of bones, floppy and wobbly and unable to hold up her bossed head, and with thin, inturned lips, no teeth and tiny, flimsy ears.

As Gladys's twin is normal, I've been reading about genetic abnormalities versus virally-induced deformities. But a friend told me about a ewe who had twin lambs, two lambs by two different rams, of different varieties – a rogue ram had jumped over the fence. She conceived at different times, the lambs were different maturities and different crosses. The difference in variety of their fathers made what had happened obvious and easy to prove. So, I wonder, Gladys could be premature even though her twin was not, and she could have had a different father – we had two rams in sequence just in case the first one had missed any ewes.

The same friend has also produced some lambs from implanted foetuses (test tube babies!) though all were inserted on the same day, there were eight days between the birth of the first and the last. Maybe little Gladys's implantation into the uterus was in some way delayed. Eight days would do it – everything is accelerated in sheep.

In truth, she probably has a genetic abnormality that should and would be fatal without my interference. While I fed her four times a day, I know I was looking for reasons to keep her, but the fact is that she is improving from day to day.

Already her posture is better, and she can hold her head up. She wriggles her tail when feeding from her mum – this is a clever signal that all is well. And she is starting to have attitude – she spits out the teat, then wants it back. She still has teddy bear ears but is starting to look more like a proper lamb.

~

Ten months later Gladys is still with us, and we have almost forgotten that she was left for dead. But, given half a chance, she grabbed life by the teat and refused to die.

She is, however, at the bottom of the pecking order. But she is fearless and curious. Or bemused. She is always last. She is the one that is missing, when one is missing – caught in the fence, or with her head stuck in a bucket, or trapped behind the gate, or stuck in the mud, or on the wrong side of the stream. As you know, if you turn a sheep upside down, it stops working. Gladys is always falling over and frightening me, making me think she is dead with her legs in the air. I turn her over and off she trots.

When a stray dog approaches their field and the young sheep run together uphill (that is the way to go), Gladys runs the other way.

She is different. She is the loner, the innocent – the vulnerable adult (just). It is my job to look after her. She is top of my list.

She is the unpromising success, the unlikely survivor, the loveable underdog, she is Kettering Town winning the FA cup.

In Nature, she is the one that would be picked off by a predator. That is her role, her niche – she's the sacrificial lamb. Sheep are Biblical.

We will keep her with her peer group until they are separated into those we keep for breeding, and those that move on as store lambs – sold at just under one year old to be fattened on another farm, and will eventually probably all go off to be slaughtered together. The worst thing for a sheep is to be separated from its peers – it is a flock animal and that is its security.

If you walk in the hills in Wales, you will occasionally encounter a strange, unpleasant smell. Your impulse, honed over thousands of generations, will be to head in the opposite direction because it is the smell of death. If you are inquisitive, you may poke around with a stick while holding your breath and you will find the decaying flesh hanging from the still-articulated skeleton. There will be lots of interesting insects and, if you look carefully (with your CSI hat on), you might notice the brambles wound around the body.

You see, brambles grab sheep. They wrap them in their tentacles, and the more the sheep thrashes and twists, the firmer it is held. It does not last long.

Don't fret – most places are so thoroughly grazed that the blackberries never get a hold. It's when the sheep sneak into places they are not allowed that the trouble starts; when they get into woods where they are not supposed to go.

Today we found Gladys in the woods. She is still testing the limits of survival. Her *baaing* drags us out of bed. She and her friend are stuck. At first sight it doesn't look impressive because, between screams for help, they have eaten all the leaves, and the thorny twine that binds them is embedded in their thick fleeces, but they cannot get away and have to be cut free, and the prickly problem painstakingly unravelled with much kicking and wriggling (thank goodness they are not fully grown).

And what is my reaction, as my hands quietly bleed? Bless her, she's got a friend!

Mothers, daughters and desperadoes

We have weaned our last year's lambs, and sorted them, with much *baaing*, lacerations, a butted head, exposure to organophosphates (or similar), unusual aggression, horse-fly attack (despite aforementioned insecticide) and general fouling with mud and excrement – and that was just me.

Now the ewes are in one field and the ram-lambs are happily in the boys' field, charging around in gangs. The ewe-lambs are very unhappily in the girls' field. This is bound to lead me to extrapolate extravagantly upon the nature of the mother–daughter bond. The ewe-lambs are screaming hysterically and throwing themselves against the double-wire fence that separates them from their mums. The mums are lying down, taking a well-earned rest and trying not to listen. You can see them clenching their teeth and staring into the middle distance.

As night falls, the *baaing* does not diminish. Shortly after 2am there is a great crescendo, and from the house I can hear the lower tones of the adult ewes joining in. I wait. It does not diminish, so I get dressed, grabbing the first garments to come to hand. The torch battery is flat. I stumble out into the starless night. It's August – where are all those shooting stars?

When I get to the source of the din, all the female sheep are gathered around a crisis, all offering an opinion. Two ewe-lambs are stuck fast between the two fences that separate lambs from mums. There is an old tree growing there that has pinned them down, resolute in its dimly-remembered hedge-duty of separation.

I climb over into the narrow wire cage, ripping my new trousers on the barbed wire, and pull the first lamb out backwards by its kicking feet and hug it tight. Then I carefully hook the lamb's flailing front limbs over the top wire of the fence avoiding the barbs more successfully than I did with my own bottom (we are talking 30 wriggling kilograms – the lamb, that is) then I heave. Amazingly, it lands like an SAS parachutist, rolling like a pro, regains its feet and in a single movement disappears into the night. The ewes are impressed.

The second lamb is huge and heavy. I apply the same technique and deliver it as a breech from the womb of the old tree. But, despite all the huffing and puffing, my strength fails me. I do not let go – I shout for my assistant ... No reply, not even from the dog. The louder I shout, the louder the sheep join in, and the denser is the silence emanating from the sleeping house.

Nothing is more motivating than having no other options, so after a little rest, I hook its feet over the top wire and with all my might I heave, and the second lamb disappears into the night.

Next morning at first light a morning chorus of ovine distress startles me from slumber (but strangely not my spouse). Exploration (slowly, as I am unusually stiff) reveals another lamb grabbed by the panicky old hedge. As I approach, the lamb butts at the base of an old fence post which, having rotted in the ground, slides to one side creating a hole and the lamb escapes.

In the light of day, the problem is clear: the newer of the two fences is fine, but the old one which it replaced, though upright, is not up to the sudden and unaccustomed onslaught of the mother–daughter bond. Hurling themselves randomly against it, the girls have found all the weak spots. It will have to be removed as soon as possible.

Twelve hours later, the last roll of liberated fencing wire is rolled towards the barn.

Remember Gladys, our 'should have been left for dead' lamb? The one with economy ears but huge determination to survive? Well, on our final trip to the barn she passes us, heading after the others, away from the scene and up the hill, far away from the mother's field, tossing her head as if to say, *We're grown up now – we're off up the top!*

My husband turns to me, 'Did you notice anything odd about those ewe-lambs?'

'No.'

'One of them seems to have testicles ...'

~

Real ram-lambs have testicles, tails, horns and bloody noses.

It's October: I know it, the sheep know it. The ewes nag me every day about moving them to the *flushing meadow*, to the best new grass which will remind their ovaries of their perennial duty. They look pitifully at their empty mineral pot and then fix me with dark quizzical eyes that ask, *When will the ram arrive?*

'I know, I know, it's not forgotten. He's booked for the ninth of November. As soon as the ram-lambs have gone to market you'll go to the good grass.'

Everything is late this year. It was such a lovely summer, autumn just crept up unnoticed. The oak trees have only just started to lose their colour, and we've been so busy with all the soft fruit, cutting timber and wasting hours trying to catch this year's remarkable ram-lambs (not to mention wrestling with a new computer, new printer, doing the tax returns, the books, the VAT, and sorting out the new solar panels and the old camper van in time to miss the very best summer in living memory).

'They've gone a bit over,' says the white-haired farmer to his grandson, looking at our crop of male lambs gathered in the far corner of their little enclosure at Aberystwyth Livestock Market. We look at the competition – pen after pen of matched, clean, docile store lambs, tails neatly docked, testicles removed at birth.

'None to compare with our tykes,' says Alan. These farmers are bound to recognise real sheep. They've got to feel nostalgic when they see these magnificent little chaps. Look at them, a band of desperadoes, decidedly *not castrated* – broken horns, two with bloody noses from fighting, not so small either!

They started the day clean and tidy but as all the other sheep in the market were trooping up and down the ramps into and out of their pristine trailers, our 18 were *making a stand.*

Turn back the clock 24 hours and we are busy trying to catch them, but are reduced to picking them off one-by-one in a makeshift trap. We are eating our lunch by the back-door, basking in the winter sunshine, with the cats and dog reclining around us.

We hear the sound of a blunt object against a galvanized trough. We stop eating and jump up, me and the galvanized husband, and we rush the 400 yards to tip-toe the last few steps under cover of the hedge and slam shut the gate, trapping one, two or three ram-lambs. After tagging them we release them into the field with the 'done-ones' and return to our empty plates – the cats are nowhere to be seen and the dog wags his tail at our return.

The rest elude us, and we retire defeated and are having a glass of wine and preparing ourselves for the ultimate humiliation – calling in the reinforcements, neighbours with dogs and long memories, to help us next morning. I have another glass of wine, 'I think I'll have one last try'.

'One more last time,' says Alan – his mantra with the children.

It is after ten. I venture out alone with the lambing torch – they have never seen the light before. I jiggle the powerful beam on the grass in front of them, they turn and run. I jiggle the light in front of the galloping posse, it stops and turns. I stand in the black night (no light pollution where we live) and direct them with my magic, jiggly beam, back and forth, slowly, little by little – down to the corner of the field and the entrance to the run that leads to the pressing pen, full of shadow and protection from the divine light. Eureka! The whole lot caught in one go – I close and tie the gate. It is nearly midnight before they are finally tagged and ready.

This morning at first light we construct an impenetrable funnel between the pressing pen and the borrowed trailer, made of metal hurdles and gates tied together with baler twine and weighted down with garden furniture – we are transferring Hannibal Lector.

We close the gate of our newly-constructed (not yet patented) sheep-machine onto the ram-lambs and we push. They compact a little. They do not advance smoothly up the ramp. They stand, their four-wheel drive engaged. They are making a stand. We push harder: nothing happens. The dog whimpers – he has no confidence in us.

Red-faced and panting, long past shouting at each other, I climb in with them. I embrace one, I pull it up the ramp and go for the next – the first is back down before me. We both try this. Alan falls over backwards, muddies and splits his trousers and breaks his wrist (probably only a little bone – he doesn't make a fuss).

Fortunately, the trailer has a full-height gate halfway along (a bulwark is always useful when transferring psychopaths). Eventually we use – *I* use – a hurdle to separate one individual from *the stand* and force him up the ramp. Then, wedging the hurdle behind me, I woman-handle him through the gate into the front of the trailer. Each time, the moment the gate opens just enough for him to see the sheep already in there, he ceases his struggling and goes peacefully. The gate only opens inwards – very well designed.

One by one we load them, some resisting more heroically than others. That's why we are late to market. 'You should get up earlier!' That's why we are not going to take them home, why my husband was raggy-arsed, why I have punk hair and khaki camouflage on my face. No one thinks to tell me until evening, why the ram-lambs have the look of *Just William* and why farmers, who are more experienced than we are, castrate their ram-lambs at birth!

'Will you take £50.50?' the auctioneer asks.

'We'd hoped for a little more', I reply to the sea of unmuddied faces (we'd studied protocol).

'I'll offer £50.80 says the handsome young dealer.

'£51?'

He looks me up and down, taking in my bedraggled state and says with an air of finality '£51.50!' The auctioneer looks for the nod which we give, and he strikes the top rail of the pen with his knobbed stick.

Another year over.

'Diana, did you count these sheep?' asks Alan, 'You see, I only make it 17!'

~

The day length is now critical, and my harvesting and squirrelling hormones are at an all-time annual high as we prepare for a long, wet winter. This, according to Iolo, who remembers many summers, has been the best ever. We know that when the rain returns it will punish us! We have too many ewe-lambs, so the excess have gone – up the hill to Deryn, who bought our lambs at market last year and was pleased to buy them privately this year. She and her husband cross their ewes with a commercial, meaty ram to produce fat-lambs for market, but need our hardy type to replace their breeding stock.

On the day we take them up, three of their number escape onto the lane. Deryn and I give chase – both ladies of a certain age. As they pass the gate to one of her fields, her own lambs stampede down to the gate to see her, led by a tame (bottle-fed) lamb. Deryn flings open the gate and lets them all out onto the road, where they mill around and sniff at our reticent three who stop in astonishment – as does the middle-aged man in the BMW, who had been giving it a burst along the lane. Deryn turns and walks confidently back to her yard, and all the lambs follow without question, including the three escapees. I think lady shepherds often do things very differently from their male counterparts (am I allowed to say that?). I am very happy that our ewe-lambs are going to be talked to (they know a little Welsh), and are not going to have to deal with shouting and sticks and snapping dogs in their new home.

Customs and a renegade lemon

The midges have gone, so at dusk I can abandon my kitchen with its bubbling cauldron of blackberries, its steeping elderberries and glugging wine jars to pick damsons to the rhythm of a pecking bird, harvesting nuts from a nearby hazel tree. There is the rustle of a squirrel filling its pouches then hitting the ground running, undulating along under the hedge then shooting up another tree. They are even busier than we are.

It is that time of year when something in the quality of light, the mist or the day-length, or the heady scent of sun-warmed blackberries in the air, turns the mind to jam. I hardly ever eat jam but nevertheless the compulsion to forage for jam jars in charity shops is irresistible. One day last week I went home with a complete stranger who thought she might have some spare jars under her sink.

As I get older my back aches a bit and my trousers get tighter, otherwise I feel much the same. But I notice that the people around me seem to ail more, and the things that fill my days are changing. A lot of the things that we do hardly merit writing about – I can't promise you a riveting account of my breast screening appointment next week.

This week I have dented the truck and got stuck in the car-wash, but I have mostly been making chutney – apple chutney. Well, I'd cleaned the house after the cider episode. The floor no longer clings hysterically to my shoes, squeaking pathetically, as I walk over it. Nor do the door handles stick to my hands, so I think *I'll fill the kitchen with vinegar fumes, taint the washing on the dryer and torture myself with chilli fingers when I remove my contact lenses!*

I can feel exceedingly green by recycling jam jars, soothing my hands in warm, soapy water, marvel at the amazing adhesiveness of modern labels and turn a blind eye (still red from the chilli) on the amount of sugar that goes in (much less than in jam). All because I read somewhere that the reason the days seem to fly past as we get older is because we don't do enough different things – distinguishing things; *that-was-the-day-I-made-the-chutney* things.

In an area like Mid Wales, where we all spend more time in natural light and so are primitively tuned by the seasons, I am not alone. I pick up the last bag of sugar from the supermarket – 'We've run out three times this month', says the lady at the check-out, 'I don't know why!'

'Bake-off!' says a young man from another planet who is queuing with his minimum-price-per-unit-of-alcohol lager.

'Jam!' says the pretty girl with the toddler who is transferring lemons from his mother's basket onto the conveyor belt.

'What a useful little boy!' says I, 'Lemons! I need lemons!' I rush off to grab two – two large, unwaxed lemons: I remember it is two because I work out the economics of it (two large ones for 40p each versus five little *economy* ones in a net for £2.00 – bastards!).

When I get back to the checkout, my husband has arrived and the lady has already put my other shopping through and is starting on the pretty girl's. I thrust my two lemons at the lady who adds them to my tally and takes my money as my husband embraces the shopping (bags cost 5p in Wales and I am forgetful and mean). We struggle out with arms full of disparate-shaped packages and bottles, all determined to escape even if perishing in the attempt.

By the time we get home they are more compliant – even the three lemons. *Three* lemons! We've only gone and stolen one of that poor girl's lemons ... And after she reminded me!

Now something very Welsh occurs.

I go to my neighbour down the lane and have a nice glass of Pinotage (that's not it). She used to work with the young man buying lager in the previous paragraphs. I recognised him, the one who was chatting to the pretty girl with the toddler (well he would, wouldn't he?). My friend rings him. He doesn't say, 'Ah yes, she's a cousin to my brother's wife', but he does know her sister and, unusually for Wales, he knows her surname which is not Jones – she doesn't live here but told him that she is visiting her Dad. Ta-da! – we've got her.

'But how did you find me', she asks somewhat anxiously. Oh dear, has she come home to Wales to escape a stalker, an abusive husband, or the Inland Revenue; has she stolen away this attractive child and gone to ground in the middle of nowhere only to be given away by a renegade lemon?

No, she remembers where she is. She relaxes. She thanks me for the lemon.

Glenys *the Lemon* – that is who she is now, in our local nomenclature. Like Dai *Bread*, the baker, who won the lottery and became Dai *Upper-crust*.

~

As the year rolls on, the grass has stopped growing, and in one chilly night the frost has felled the bracken that we battled all summer. Every tussock of pasture has been grazed to a smooth hump (where the sheep can't reach, the rabbits have obliged). The wind has denuded all but the youngest oaks on the hill and the gales have scourged the yard clean. All is cold and strangely tidy. Once they have fed the sheep, cracked the ice on the troughs and checked to see if any early frog spawn has appeared in the pond, farmers can sit for a moment by the wood-burners in their warm, marmalade-scented kitchens and think about their position – their place in the world …

'The measure of a man's importance is the size and number of his woodpiles.'

I was told this fact many years ago in rural France. It made a great impression, so contradictory was it to the progressive philosophies of my young French friends that I found it oddly reassuring, and still do. We have woodpiles – burning wood when you have lots of trees is great, but trees need cutting down and they don't go quietly. They have a lot of stored energy and can lash out ferociously. They need logging and drying, and dry wood burns amazingly quickly, so you need loads – and plenty of room for storage. We have an old barn, 30 feet by 20 feet, already full of timber.

Last back end (as they say in Lancashire), we culled a Leylandii hedge, grown 40 feet high in a blink of Mother Nature's eye. But when the exalting roar of the chainsaw had stalled for the last time, we were left with a daunting amount of timber, a mountain where our new workshop is waiting to be built. We cut off the branches and burned the brushwood. The trunks lay where they fell, until last week when we were taken in hand.

Not by the Forestry Commission or the satellite snoopers of Rural Affairs, Wales (which has nothing to do with illicit romance in the hills). It wasn't even our very grown-up off-spring who, though they never tidied their rooms, now worry about the *state* of their decrepit parents. No, it was a young neighbour (well, relatively young) who knows that all we need is a tiny push, a little encouragement.

'I'll come and help you on Tuesday. I've nothing much on this time of the year. I'll be with you at midday.'

We refuse, we protest, we are tempted. We eat at midday, we say, he'll have to have lunch. 'Will there be meat?' he asks.

'Yes.'

He accepts. We capitulate. It is arranged. That's how things are done here. In the intervening few days, we have got on with what we should have been doing for months.

Today is Tuesday and we have started two new woodpiles, and now something strange is happening. Tree trunks scud over the ground, whizz through the air, crash into trailers, flatten the saw trestle and just about spifflicate two pensioners, temporarily under vigorous new management.

Chainsaws start willingly, and logs march to the music of the Sorcerer's Apprentice, jumping happily onto the new woodpiles.

By evening, by some marvel of effort and teamwork (mainly one man's effort – we helped as hard as we could, and tried not to get in the way), we have uncovered the bare earth where our new workshop is to be sited.

It's a miracle. Just another of the miracles of living here – Thank you Gareth!

~

Once a week, during the winter, we go into town to meet some friends and have a pint (I have two halves – I'm a lady). It's a funny old pub. The landlady has been there 50 years – her longevity is probably because she doesn't drink or smoke. Nothing has changed in that time. On the wall are pinned all the postcards that her regulars have sent back home over the years. You can plot the changes in the nation's holidaying habits. She shows us one she has just received courtesy of Dai, from Thailand – he's been back a fortnight. She pins it back up between a sepia image of the harbour at Aberaeron and that of two little girls by Lucy Mable Atwell, plump and conspiratorial.

As I stand, waiting for my beer by the little hatch that is the bar of the cosy front room of the pub, Alan chats, through the hole, to the men standing on the other side in the even smaller public bar. I look down at the blazing hearth and I see them, they are by the fire with the tips of their middle fingers just touching, their thumbs pointing out into the room and the little fingers reaching towards the heat.

The gloves are grey, hand-knitted in a plain stitch without pattern. Sure enough, Norman, their owner, is sitting in his corner, motionless, immaculate, his grey hair now corrected from any dishevelment caused by the wind outside, his parting pin-point accurate, his comb restored to the top pocket of his grey tweed jacket, his overcoat hanging from the only hanger on the hooks by the door. He has brushed the single speck from its shoulder with his then still-gloved hand.

His back is straight and rests against the upholstered bench that runs around the wall of the front bar. His legs are stretched out in front of him, beige trouser-creases crossing below the knee, polished brown shoes reflecting the fairy lights from the bar. The evening paper is neatly folded on the shelf beneath the table – he does not touch it. He looks out into the room without expression, a glass of orange squash, obtained without negotiation and paid for with exactly the right money, sits untouched on the table in front of him. It will remain there until just before he leaves, when he will drink it down in one and return the empty glass to the bar. He will weave unseen through the people and leave, but we will notice that his gloves are gone.

Norman has a story – everyone has a story. The more controlled a man's world is, the more bizarre is the erruption when it comes – and come it always does.

I will not tell you Norman's story, for the one I have may not be true (and anyway, you will probably have heard it before).

~

Winter is no longer wet and soggy or scoured and dry: where we live it is suddenly white and crisp.

Today we set off to inspect the moors above our home on the untreated roads. There is an amber alert for heavy snow overnight, and cautious farmers are driving their sheep to land nearer home.

Like us, the sheep are slithering a bit but seem pleased to be heading home – unlike Alan, who wants a bit more of an adventure. So, the following day we set off with diminished responsibility to test the roads and our new off-road tyres.

So far, so good.

Up the cwm we go and into town through six inches of powdery snow. All the twigs of all the branches of the oak trees that meet over the lane, carry an icing of snow so the whole world is white, and where the sun strikes the tops of the trees, lumps drop onto the truck, snowballing us.

In town, the Co-op's shelves are almost empty – their lorry is missing, presumed lost (come to think of it, we saw a lorry stuck on the hill). We get our pills and enough ginger wine to last until spring, and speed homeward.

We think we'll avoid our dangerously-steep cwm with its sheer drop one side and all the inconveniently placed oak trees, unyielding in a slippery situation.

We go the other way – we are trying to be responsible.

It is odd that there are no tyre marks into our turning, just beautiful virgin snow – powder, if you are a skier, about a foot deep up here. We chose this route as it has steep banks (quite a lot of roads here are narrow, tarmac strips suspended between precipice and ditch). This way, we can see which way the road goes. As we drive higher, we can see the drifting: ridges of white dunes cross our path from bank to bank, deeper and deeper as we go higher and higher. Now we remember seeing the drifting starting two days ago, before last night's heavy dump. There are no houses up here and no lights – just white drifting snow and wind.

'Shall we go back?'

'Not yet!'

The technique is to drive as hard as you can until the drift stops you, then reverse and do it again. Each time we get a little further using our makeshift bulldozer: back and forth, higher and higher, deeper and deeper!

Amazingly, we reach the crest, and it becomes slightly easier as we descend into the dip – into the unknown. We turn at the bottom and can see the tracks where a quad-bike, coming the other way, has given up and turned for home. That cheers me up. I am being very quiet and brave! We follow in its tracks.

Now we have about half a mile of a straight, steep rise which Alan takes at speed, relatively, drifting and sliding, sometimes almost travelling sideways but keeping going.

Then suddenly we have crested the hill and can feel the road, solid again, under our wheels.

Hot tarmac, oily rags and paraffin

Our responsible friend, Aled, is drowsy with counting sheep, coming up the valley every day after a day's work, to be jostled by our impatient flock. Because this is the time of year that we go gallivanting.

But first, the Responsible Friend has noticed a hazard in the field in which we feed our precious ewes. We have to fill in the 70-metre trench that we dug for the solar panel cable. I use the word *trench* appropriately, as torrential rain renders it a living memorial to life on the Somme in the First World War.

The week before we leave for the West, we slither and shiver, often up to our knees in mud. Alan's relationship with little Digger, now we are past the honeymoon stage, is tested almost to destruction (not a bad thing as they were getting far too comfortable in each other's company). Her electrics are labile (common at her age), making her temperamental and unreliable so that she often refuses to work at all and sits facing the prevailing storm with her broken windshield, getting her seat wet (but then we all get wet seats).

West-ward ho we go in our scruffy white camper van, not a frontier wagon, like the film, but the modern equivalent of the genuine Gypsy caravan, evocative of prejudices dating back centuries. It has only recently come out of retirement, rescued from our barn, emptied of animal feed sacks, and given a cursory vacuum clean. We've been busy for the last ten years and, as the only completely rat-proof container, it's been busy too, minding sheep-nuts and sheltering privileged spiders. Now it's time for a rebirth, an adventure, a pilgrimage, a journey.

Our camper van is a peripatetic, psychological assessment tool. It is an indicator of people's preconceptions. Driving it is chastening, like going round a supermarket in a wheelchair ('Ah bless!' said the cashier to me as I tried to pay for my shopping shortly after spraining my ankle). When we drive it, gates close, barriers come down, appeased only by the roundness of our vowels and the friendliness of our dog.

To us, it is cost-effective. It is a warm and comfortable bed in a light and airy ex-commercial, VW high-top Transporter – old, high mileage, empathic with no fancy electronics to go wrong and no frilly curtains. It is insulated, ventilated, has running water (usually), a fridge, a cooker and a loo. It smells of oily rags and dog, but they are our oily rags and our dog. When parked overnight in a municipal car park, it is just another white van and goes unnoticed. Best of all, no-one cuts us up on roundabouts; we look as if we mean business (even if it is the scrap-business). Unlike B&B, there is no fuss. You don't have to be endlessly polite to your host or worry about the dog barking if anyone uses the bathroom. A huge man does not stand over you while you force down the largest full-English breakfast in Cornwall, telling you about his most recent coronary. No-one sniffs under the door to see if you are smoking or charges you £15 extra for the dog who is on a diet and doesn't need the sausages either. When you get tired, you can just stop and have a sleep – it is perfect. You can drive to the beach in your pyjamas and walk the dog while your spouse snores on.

~

When we made this journey last summer, we picked our way over the mountain pass to South Wales at walking pace. We avoided the hundreds of road-runners who were jogging up on the hottest day of the year so far. We gave them a wide berth to allow for heat-exhaustion-wobble, weave and collapse while also avoiding the pulses of road-racers on two wheels coming the other way (lucky to have two lanes – this is Wales) – pelotons of cyclists, who had just crested the summit, heads down and were hurtling in squadrons, turbocharged with huge potential energy and suicidal intent, lemming-like, towards Brecon.

We glanced at the stunning scenery and at the idyllic path on the other side of the valley, made for walkers, and wonder why it is that so many humans have become addicted to hot tarmac. Our musing is ended abruptly by the thud of a discarded plastic bottle flung, elite-runner style, against our windscreen by a mature but plucky lady with exceptional BMI and poor aim, probably due to chafing.

As we eventually sped away from the last; or rather, the first of the runners and the last of the cyclists, the bikers start to overtake us, flashing past at every opportunity, like when we slow down to turn right. I have a horror of killing a biker, and they come to Wales in huge migrations at holiday times: Hell's Angels and 1950s re-enactors on vintage Nortons with side-cars. Even an indomitable band of ladies, several with L-plates, on vintage Honda '50s', almost grinding to wobbly halts on the hills – though that was on the A30 high-speed dual carriageway in Cornwall.

~

Now noisier than we remember, Camper Van discourages unnecessary conversation. We nod at Glastonbury Tor as we chug past. It is promised to friend-Silvia for her bucket-list trip to the festival, but we have enough mud in our everyday lives and spend our time trying to avoid crowds. It hasn't lost its charisma – landowners pale at our approach.

We make it to Cornwall again without fatality.

Turmoil comes in waves, and these times of upheaval are our most creative – this is what I tell the children (it is no comfort to them).

It is Christmastime and our five grown-up children (now that's a strange concept) and their partners all seem to be facing new challenges. Four now have various commitments in the West Country – work, homes, other family and friends. So, to make it easier to all be together this year, Mum and Dad (and Pedro) have come West. Christmas is packed in Tupperware containers and here we are at The Gables, a rental house in Tywardreath, Cornwall.

Cornwall is dry and comfortable. Camellias bloom in gardens on Christmas Eve. There are ragged robins in the hedgerows as shop assistants glower at befuddled consumers and cars queue to enter and leave the supermarket car parks of the peninsula. Sensible folk walk their wet dogs on nearby Par Sands where the China Clay factory breathes steam into the clear chilly air.

Carols are sung in the pub – an excellent compromise and a solution to the 'middle-class-at-prayer', who cannot shrug off the prejudice that going to Midnight Mass after an evening at the pub leads to embarrassing behaviour, even from those whom one thought one could trust (and no, it isn't really a birthday party!)

The sun comes out on Christmas Morning. We toast the Messiah with bubbly on the patio. The silvery sun makes it imperative to get out and make the most of the short days.

Festive meals are served for various permutations of family and friends.

The wind changes, coming in from the north-west. We pack up the leftovers and drive home avoiding *Bannau Brycheiniog*, the Brecon Beacons, but not the traffic, to arrive home as the cold freezes the first dusting of snow into a crisp sugar coating over everything.

On the way home I tell Alan about my best-ever Christmas, ironically the year before I met him.

The year 1992 was Queen Elizabeth II's *annus horribilis*, when her life was turned upside down and the outside world came crashing in – we all have to suffer a re-set occasionally. Our own personal crisis was two years later – no-one is exempt. As Christmas approached (the first without my children's father), we knew that it would be terrible. That day so laced with expectation – and us with our open sores.

One of my daughters pointed out that it had been a very bad year for family breakdown. We thought about all the people we knew who were struggling that year: my friend in a similar situation; my cousin whose partner of 20 years had died after a harrowing illness, leaving her with all manner of difficulties; the doctor I'd worked with, loved by his community, then in his late-sixties, cast adrift after 45 years, on the day he retired, by his wife, right-hand and practice manager (married to the job, not him, apparently); the bereaved and abandoned children (age it seems is no protection); the Granny, come to live with her daughter just as the daughter's family falls apart.

That's who we were that Christmas. For the first time we realised that there were other people who were alone and unhappy – not freaks, but people like us. Not all strictly alone;

several of us had children. We felt bad that we'd never thought of it before. Bad things do happen to everyone, but good things can come out of bad.

On Christmas Day, four women, all supreme in their own kitchens, their own Christmases, stood stirring around the central hob, with no vying for dominance. None of that family nonsense – we stirred as one. Tom, my old Senior Partner, attended the turkey.

When the sprouts boiled over despite eight eyes watching them, we laughed until the tears ran down our faces. It was the first time that had happened to me (the tears of joy, I mean) for years and years. But, you know, it was to happen more and more.

The children watched their new, almost grown-up friends, almost cousins, showing off their circus skills.

They were fire-eating in the garden – spraying flame, like dragons, from their mouths, spewing blazing breath diagonally into the air without setting any of the younger ones alight. Not physically anyway. No-one was injured having goes on the unicycle, not even on the tall one; or concussed by their new diabolos. Glasses in hand, we watched their antics through rainbows made in the winter sunshine by the giant bubbles they blew.

After lunch, we lolled on the sofas and on cushions on the floor to watch the Queen's speech, in the euphoria of full stomachs and alcohol, moulded to each other, inspecting singed hair and smelling slightly of paraffin, in comfortable congestion, like a pride of circus lions.

That was how we had our best-ever Christmas.

Aerial dog fights – Roger falls to earth

We are not in a war zone, but over the undulating landscape of Mid Wales, fighter aircraft of the Royal Air Force rend the sky and intertwine their parabolas as they pass behind the hills to re-emerge and cross, one with the other, with micro-second clearance. They travel in pairs, weaving like mating dragonflies on amphetamine, never quite making contact, thankfully – so far.

They use this area for low-level training (I don't think it's a secret) and use our house as a landmark. Or perhaps we are located exactly on the intersection of the invisible lines of the virtual grid that is projected onto the land by a NASA satellite (the eyes in the sky). When we were slating our new roof, the eyes in the sky were obviously interested, sending fighters to make pass after pass over our house, lower and lower in the sky, trying to topple the large khaki penguin, wrapped up against the elements (it was winter and we do everything late). Were we part of a secret military exercise? A pretend enemy missile installation under construction? Subject to constant aerial monitoring, and due for annihilation when we fixed the last ridge tile? Or was the intelligence officer just keen on DIY, trying to see how we feathered and leaded the valley of our new roof?

Anyway, we enjoyed the attention.

We're not paranoid, not even when a massive Hercules transport plane hoves over the horizon (which, in these hills, can be just yards ahead). Motorists on the mountain road swerve to avoid the huge alien craft that rears up as they approach the crest of a hill.

The remains of a fuel tank from such a low-flying plane,

was in our barn for years after the last war, jettisoned by a pilot who misjudged the height of our hill, and quickly squirrelled away by conspirators to fill the oil lamps of this valley for a generation – or so they say.

The aerial activity recently has been more pastoral. The crows that roost and build their nests in the wood do not like the buzzards, nor are they very keen on the red kites that swoop down from great altitude to pick up the remains of pheasant carcasses left on the hillside for them by this lazy farmer's wife who is tired of making soup.

The buzzards are ever-present, mewing to each other and circling above the trees and crossing the valley. The crows are intelligent and social creatures and resent this invasion of their air-space, so have formed an air force of their own. They climb up high in ones and twos and swoop down on the buzzard from above and behind, and the buzzard will twist and roll to face the enemy with his talons outstretched, and they will engage and drop and spin in the most aeronautically alarming way – a real dogfight.

They recover and the buzzard continues to beat his Herculean way across the field of combat as the crows re-form to attack again.

It's hard not to sympathise with the plucky crows, especially after the chicken incident. Imagine our delight when a great bird of prey alights just under our bedroom window to consume its prey. We are honoured and watch and wait, enthralled, to photograph its every move, and later rush out to examine the spot – only to discover the remains of our last bantam hen!

In my excitement and ignorance, I failed to identify that this chicken-killer was in fact a goshawk, an aristocrat among hawks, but still not entitled to eat our hen. This is brought to my attention years later when I am showing the photographs to a friend. Too late to ameliorate my anger.

~

I hardly notice the tap at the door – most neighbours just walk in. Whoever it is, rings the ship's bell that hangs by the door to call in the worker for lunch.

I throw open the door.

'Sorry to bother you but there's a man in the ditch who says he's got a broken leg.'

'Who is it?' I ask the worried faces at the kitchen door – one is a workman adjusting our solar panels and the other is a white-faced lad with beads of perspiration on his fuzzy upper lip.

'Can't remember – he did say his name,' says the lad, 'I nearly ran him over! He told me to get you!'

'Is it Roger?' I ask over my shoulder, running to the gate.

'That's it! That's his name!'

Roger is our nearest neighbour. He is lying face-down in the narrow, shallow gully that runs down between his house and the road. He is darkly-dressed, and mud-splattered, and still wearing the world-weary cricket hat he had on a couple of hours earlier when he had been in our kitchen drinking coffee. He is perfectly camouflaged, but the other workman is standing guard to make sure no one else runs him over.

'What *have* you done?'

Roger had skidded on the slippery ramp to his cabin and heard his ankle crunch and snap. He had called and called, but we were digging with the noisy digger, and no one heard.

He tried to adjust his right foot into a walking position, felt faint and thought better of it. He shouted some more, and no one heard. His wife is out, and he doesn't know when she will be back. It started to rain, and the sun had sunk behind the tall trees, and he was getting chilly, so he set off to crawl the 50 yards through the long, wet grass to the road. He was on his way, commando-style, down the ditch towards his front door and a telephone, when the man delivering our cable (fortunately young and on-the-ball) narrowly missed him and got out to investigate.

While the workman calls an ambulance, I wrap him in roof-insulating foil and carefully unlace his boot, it does not seem to hurt him too much.

'Perhaps it's just a sprain?' he says.

'Perhaps it is. Can you roll over, and we'll have a proper look.' He rolls over and his booted foot flops into a strangely unnatural posture.

'Whoops! Roll back Rog.'

We remove his boot with the foot pointing in its normal direction – aided by gravity. It is warm and not all that tender or bruised, and I can feel several pulses. We wrap it up to keep it warm and await the ambulance. Several vehicles come along – all stop; the drivers get out and join in. One wraps his fluorescent coat into a bundle and puts it under Roger's head. Someone else gives him a lighted cigarette. A police van arrives, the first we have ever seen in these parts. It is only passing through, but the driver waits patiently behind the logjam of other vehicles and chats. He does not get out.

Roger is feeling quite warm and becoming positively effusive. I've noticed this before – something to do with adrenaline, I think. People can seem at their very best when they are quite near to their very worst. It is probably the secret of most heroism ... It won't last!

'Here it comes!' the look-out shouts, and a big yellow *ambwlans* sweeps onto the scene – we are chastised for the smoking. Roger is loaded and someone slips his rolling tobacco and papers into his soggy pocket. The doors of the ambulance are closed. I am sent to find some dry clothes – not easy in someone else's home. I do the best I can. I am then dispatched to find his medication and to pack an overnight bag.

Meanwhile his wife arrives, shocked by the sight of the ambulance and surprised to find the paramedic dressing her husband in drag. She retrieves her clothes indignantly and makes haste to procure more macho garb; she also manages to find his pills. And off he goes.

Later that night, his wife returns from the hospital at Aberystwyth. He is to have his ankle surgically pinned the following day on the other side of the Principality. She has already washed his wet clothes and pegged them out.

'What do you think I should do with his tobacco?'

'He won't be able to get out to smoke.'

'No, I know that – it's been through the washing machine.' We both peer into the packet. 'It's only a bit damp.'

Two days later we visit him in the orthopaedic ward at Gobowen, only 50 miles away. Thank goodness he wasn't transferred to Cardiff (that would be a 200-mile round-trip). He is sitting up in bed, very clean and pink and still in high spirits.

I notice the T-shirt he is wearing has pretty little flowers embroidered around the neck which, now I come to think of it, I vaguely remember bundling into a rucksack. I don't mention it.

Hiraeth

Hiraeth is one of the most important words in the Welsh language, yet without an exact equivalent in English – that says it all really.

An Englishman would say homesickness, a negative feeling that unsettles you and stops you doing your job or studying properly. But in the Celtic vernacular, *hiraeth* is a sense of incompleteness tinged with longing. It embodies the yearning for the spirit, the beauty of the landscape and the belonging to a place.

It is that feeling we have at dusk, in the bluebell wood. It is love. It is God. It is home.

It is not a one-way thing, either. Our home seems to have its own feelings and longings of its own.

It is forever making its presence felt, whether via its climate or its wildlife, or by knocking over trees, felling elderly gents, or blocking culverts. The place has its own agenda, its own friends and relations (they often visit unannounced for tea), and it is quite possessive.

This morning it's Sunday, and we're celebrating Communion – coffee and biscuits with Roger. He hobbled over on his crutches and now sits with his bad foot up on a chair. There is a knock at the door. Outside, in the drizzle, is a young woman we have never seen before. She is waving a long cardboard tube.

'You don't know me, but I've come on an adventure!'

She isn't after our souls or even trying to sell us something.

She is just another one of the people our cottage-holding has sent for (it's happened before).

Deborah is an artist and buys strange things that inspire her at auctions. She cuts them up and stitches them. She bought the map of our place – a 1901 Ordnance Survey – at a sale in Leek, Staffordshire. She bought it years ago but could never quite bring herself to cut it up. It was personal to the house, you see – it has all the field names pencilled in, in Welsh, and even has the new well marked (circa 1980).

She couldn't use it, and was passing within ten miles, so has brought the map home. She can't explain it and feels it is rather an odd thing to want to do. But we don't – we know our home. It doesn't like to let go of things or people. So, she has a cup of coffee, and we will hang the map, once framed, next to the horse brasses, the dresser, the polished pump-nozzle, the wood-wormed rake and the photographs of past residents and their New-World descendants, who have visited – all things that this sentimental old homestead has collected or reassembled since its original scant contents were all dispersed at a farm sale in 2005.

I will enjoy showing it to Myfanwy when she next visits the old place – she will remember it. She will do what she does: drink tea and gaze fondly at the drying rack that swings above our inglenook. Her husband of 70 years made it for her when they were courting because she complained that she couldn't get the washing dry and couldn't come out with him until she had finished the ironing. It was made from timber scavenged from the farm and held together with love. As she looks at it, she remembers it all – the black kettle steaming on the old range, the pitching floor of tightly packed pebbles set into the earth that had to be swept and mopped every day – until her wedding day when she was shocked to be told that she didn't have to do it before she set off for the chapel. We found the iron in the field with a metal detector – it sits on our wood-burner.

Myfanwy came to live here in her early teens, on the day she found her mother, lifeless and covered in blood, up on the hill above this farm. Her mother lay dead on the wet grass in a field behind the house in which they lived – Myfanwy and her three little brothers and her mother and her terrifying, alcoholic father. She cried and shook her mother's body, but she would not wake; she could not wake. The discharged shot gun lay on the grass beside her.

This was how Myfanwy described it to me. Others, she said, remember it differently. She recalls how her father took her by the arm and pulled her, dragging her down to the grandparents' farm, this farm, her brothers following, frightened and crying. 'Here', he shouted at his parents-in-law, 'she's only gone and shot herself! You can have these!'

They did. But not without a fight. It wasn't easy for the grandparents in a subsistence cottage-holding with a little parlour and two bedrooms, with six of their own children still living at home, the youngest backward, and a lodger. And now four more mouths to feed – three of them riotous little boys. Just as things had been starting to ease; their sons getting strong enough to be useful and the elder daughters had found places as domestics in other, bigger, farms. Suddenly, all this to face. Hard on Myfanwy too. 'Twelve of us and a lodger in this little place.' But much better than the alternative.

A photograph of the time shows a world-weary old couple. He: thin and craggy, staring across the years, defiantly, with a tall, black, felt hat and long, drooping moustache, like a scruffy frontiersman. She: swathed in layers of worn, homespun clothes, her straggly white hair pulled into a bun, standing in heavy boots in the muddy yard. Their image is timeless, give or take 200 years, but the last of their many children died only since we have known the place.

This couple, forged in poverty, did something that was almost unheard of in their day and in this location – they took their son-in-law to court. Whatever their suspicions about the circumstances of their daughter's death (and you can be sure they had severe misgivings but no proof) they were not going to allow him to just walk away. They were driven, too, by the practicalities of life. They sued him for maintenance for the children and they won. Not a great deal, and in no way compensation for their loss or their sacrifice, but a triumph, nevertheless.

The three little boys have been back to visit the old place, to reminisce about their childhood, their pranks and their high spirits. Their Grandad, Taid, strict, harsh even, but fair.

Happy memories.

~

There is a man who is remembered in Llanidloes from a time when the town was in its footballing heyday. I have been reading about him in the archives of the local paper, and today I have been talking to one of his younger sons, now an elderly man.

It was not the football that was the most memorable thing about him. Although between 1932 and 1937 he was capped for Wales seven times, and played inspiring football for Llanidloes, which was the talk of the town and of the foundry where he and so many of the local men worked. From Monday to Wednesday the talk was all about the previous Saturday's match, and from Thursday to Saturday midday (the end of the working week then), the talk was all about the next match. He raised morale and gave a focal interest to the townsfolk at a time of recession when there were not a great many things to do. He was quick-witted and fleet of foot, and arguably the best footballer that the town has ever produced – but football followers do love to argue.

There was another side to this man and that was what one might call 'field-sports'. Gurra Mills was also a notorious poacher. Although some landowners might have felt quite honoured to have this popular celebrity catch their fish or shoot their rabbits, Gurra was always discreet. Though it was public knowledge that, after a trial for Arsenal in the early '30s (after which he was nicknamed 'Gunner', later corrupted to 'Gurra'), he returned to the town, explaining that he had turned down the offer of a professional career in the game because there were no fish in the Thames and no rabbits in Woolwich!

He had trials for Wellington, Shrewsbury and Swansea. But, when it came to it, *hiraeth* drew him home. He was not prepared to leave the life he loved in Mid Wales.

At times, Gurra's interests converged – as when he was touring Scandinavia with the Wales football team, during which they were feted by King Gustav, who awarded them all commemorative medals. Perhaps it was just before this or some other municipal reception when half the team went missing. Where were they? Gurra had them all out on the river, looking at the fish!

These were the 1930s, and Gurra and his wife Sybil had a growing family. Life wasn't easy, even for a local hero. He grew potatoes at a local farm in Llandinam, where he also shot pigeons from a hide made of straw under a great oak. He would help out with threshing on other local farms.

Setting off, well before dawn on his bicycle (he had crashed his motorbike), he would leave his bike at Glan-feinion, the farm in Llandinam, and walk back through the waters of the Severn with a miner's carbide lamp in one hand and a gaff in the other (a gaff is a pole with a large, barbed hook on the end). By the time he got back to Llanidloes, his flap-topped wicker basket would be nearly full of trout. The larger ones might be sold to the baker on Victoria Avenue, and the smaller ones cooked for the children's breakfast. In the early days he had dogs (whippets or lurchers), and he went after rabbits with ferrets and nets. He had a .22 rifle with telescopic sights, and he knew exactly where to go. But even his children didn't know where. He was always careful, always guarded.

The pot was never empty. He shot snipe, woodcock and partridge, black grouse from the Gorn Hill and the odd wandering pheasant. His children remember the pungent smell of fish roe, cooked into a paste and smeared into beer-bottle tops to toss into the rivers to attract the fish. And they recall the scramble to burn the heads and tails if there was a knock at the door. Neighbours would wake to find a salmon hanging in their coal shed, and the coal-man said he never knew what he would find hanging in Gurra's coal shed.

Gurra was never caught poaching – unlike another man, up in court for having a gun without a licence. Gurra would tell the tale of how the news of his conviction reached local drinkers at The Stag Inn: 'Found a pound and his gun was constipated!'

Just as today, when the movements of the traffic wardens in Llanidloes are magically transmitted around the town, so in the early part of the century the movements of the *Hafren* keepers on the River Severn were known. They drank in The

Lion at Llandinam and, once safely installed with their pints, word would spread up and down the river that it was safe to fish. On the coldest nights, when the keepers clustered around the big fire in the taproom of The Lion, Gurra would be fishing, returning at dawn, soaked and at least once, with his trousers frozen solid.

Little wonder then that by 1940 Gurra's footballing career had dwindled. Not because of the Second World War though – he had been having night sweats for months and had contracted tuberculosis, together with three of his children. He was hospitalised in Machynlleth for 18 months, losing three-quarters of his lung function. Two of his sons were in hospital in Llandrindod Wells and his younger daughter in Anglesey.

Gurra will be remembered as a raconteur and wit, though there are many stories that he would not have wanted in print – he was always discreet. Later in life he was to be seen around the town in his shirt sleeves, flat cap and braces, baler twine round the top of his trousers, perhaps collecting the swill from the Trewythen Hotel or The Angel for the pigs that he was allowed to keep on some land at the foundry, rented for the occasional half-pig.

In that same bucket with the lid, he would carry fish to leave in the shed of a neighbour – nothing said, but everyone knew it had come from Gurra.

In the old days, the town siren was tested every day at 11 o'clock. On one August bank-holiday Monday the foundry was closed, and Gurra and a crowd of other men were chatting under the old market hall, waiting for the pubs to open. A coach from the Midlands stopped and asked for directions to *Rhaeadr* (Rhayader), but the conversation with the driver was interrupted by the siren. 'What's that?' asked the driver.

'Opening time!' said Gurra.

Well into retirement, Gurra was the conscientious agent for a football pools company, going round the pubs on a Thursday evening, collecting the money and the forms. He wore a leather shoulder bag under his coat, in which to put the money, and if anyone got too close to this satchel it would growl ominously: Judy, his little dog (a Miniature Pinscher, like a tiny Doberman) just fitted into his poacher's pocket!

Throughout Gurra's life, there was one story that he repeated to his children, the events of which had obviously made a big impression on him. It was when he had been helping with the threshing at a struggling farm in the '30s, with no prospect of payment other than a square meal at lunchtime. As they arrived, Gurra noticed that the farmer's dog was nowhere to be seen. 'Where's your dog?' he asked, but the farmer only shrugged, looked away and changed the subject. At midday, lunch was served – a large stew of white meat, enough for everyone. This tale of Gurra's showed just how difficult life was in those days of the Depression – for the rest of his life Gurra remained convinced that they had, that day, eaten the farmer's dog, and I repeat the story to put Gurra's poaching into the context of those hard times.

All the misgivings that we might have about the morals and ecological effects of poaching dissolve, as Gurra's own story about the farmer's dog reaches to us down the years and tells us about the real hardships of those days.

Love story

'You make me so angry!' I bellow. I am stamping my feet. We've been together more than 20 years – when we met our joint age was already over 100. Now he is standing, unsteadily, on top of a curved and slippery plastic fuel tank which, in turn, stands on a concrete plinth as tall as a man. 'I can't turn my back for a moment!' In his hands is a large but silent chainsaw. All around a hailstorm rages. He moves his feet a little: they crunch, he wobbles, he laughs.

'It's okay,' he says, 'It's quite stable.'

'It's slippery. It's round. It's wet plastic.'

There is a tree suspended uncertainly above his head. It spans the space between its base, where it normally stands on top of the bank behind our house, and the house, on whose corner it now rests. It was blown over in yesterday's storm.

'Come down! If you fall, you'll break your femur, or your neck, and by the time I get you to hospital you will have bled to death.' (I'm always mindful of his anticoagulant status.)

'Don't fuss.'

'Please come down.'

He pulls the starter and raises the roaring saw above his head with both hands, showering me with sawdust as I look up, both arms raised in supplication or ready to catch him and have my head chopped off. The tree above wavers – whether 'twas better to knock off a few more tiles or knock the old man off his perch.

'Pull the rope,' the old man shouts. I pull the rope. It is attached to the tree (now that's a first – he generally attaches the rope *after* he cuts). One of us groans, or it might have been the tree clutching at the guttering. The tree keeps hold but the end of the gutter comes away and the pipe sags, soaking me with ice-cold water and wet leaves.

'Come down – PLEASE!' He climbs down with surprising ease, having brought the wobbly step ladder out of the airing cupboard and placed it against the back of the tank.

I am now thinking on my feet. I pick up the long, aluminium ladder that is lying nearby and fling it against the bank. 'Look! Climb up that – you won't slip, and you won't fall so far. Worst case scenario – you'll roll, like a wicket keeper. Cut the tree near its base at the top of the bank and I'll pull.' He does. And I do. And the tree lets go of the house and falls to the ground, bouncing on the concrete. He chops it up where it lies, and I pull the logs and branches out of the way.

I look up and he is back up the bank, silhouetted against the sky, gleefully rocking a large rotten tree trunk back and forth.

'Look at this one.'

'Oh, Alan!'

'There, you see, it was alright, wasn't it?'

~

As the sun appears over the hill, the whole area is bathed in amber light reflected from the dying bracken.

The woods are glowing with new colours. Dew on spider -silk drapes the dead stalks of yarrow in gossamer, and polishes the red tips of the mellowing bramble. Even the dead wood on the compost heap is looking its best.

The year rolls on unremittingly. Minty-ness rises from the damp litter of fallen leaves.

We have enjoyed a long and beautiful Autumn. The beech

woods have been aflame, and the more sober oaks have held on to their russet leaves until just a few days ago.

But now suddenly, in one night, everything is changed.

The sky has cleared, and the temperature has plummeted. At night the stars in our black night are stunning, and the all-day frost in the hill's shade makes the morning seem moonlit. You can see the cold and smell the cleanness of the air (and trip over a frozen molehill).

Alan has just informed me that he thinks he is probably 'winding down to die'.

Only as the low winter sun peeps over the hill at noon and stretches over the ground, does the colour return for a moment – otherwise all life seems to be drawing back into the cold ground.

I have just told Gareth what Alan said. Despite the age difference, he is Alan's closest friend. I suppose it is a father–son sort of relationship. They seem to relish each other's company. We are standing by the gate by the barn. I have lured him out of Alan's earshot, and it all comes tumbling out of my mouth with a blast of terror from my eyes. He doesn't answer – what can he say? He returns my gaze and I feel better somehow – like a pressure cooker (do you remember those?). It has blown its weight off and is spraying scalding steam at the ceiling and the pressure in my head is gradually reducing. We go back to the house without having looked at whatever I said had to be seen.

Next day we are at the doctors. Alan is nearly 78, he smokes heavily and unapologetically but has never had a day's illness in his life (apart from atrial fibrillation which is not a problem, and shingles, a little while back). He has no localising symptoms but is sent for a chest X-ray and a blood test.

Two days later the doctor phones to say that it is all clear.

I am relieved. I don't even mind the label of *Worried Well*, if only by proxy.

I am intuitive though.

And not stupid ...

Sunday craik

God spends quite a lot of time in the pub. Sometimes he's there when he's supposed to be in chapel – that's what I say to friends who ask where we are bound on a Sunday afternoon when the good folk are heading down the valley to the chapel.

Today there is an added incentive (for the pub, not the chapel): Liverpool are playing Man City in the League Cup. It's not that Alan supports Man City (he would warm to anyone in competition with Manchester United – it's an underdog thing). The landlord of our local is an avid Liverpool supporter, which adds to the fun enormously.

'Can we have the Rugby on?' asks Alan as we arrive.

'There isn't any!' snaps the landlord.

'Wasps are playing against Quins on BT Sport.'

'Can't afford BT Sport with the pitiful amount you drink!'

'Do Wasps have a 'B' team?' asks Ikey, '*Bee* team', he repeats. A muffled titter runs around the room as I am reminded of the one terrible joke that my Dad knew – no point repeating it, no one will laugh, they never did!

The landlord asserts himself by switching on the commentary. That way he can follow the action despite all the distractions we can throw at him, like the full glass of Stella

I knock across the domino table due to the excitement of a penalty – it misses Alan almost completely.

As the match progresses, the joy of winding up the landlord is irresistible. People who normally have no interest in football *whoop* with every Man City tackle and berate the ref for every decision that favours Liverpool, carried on a wave of affectionate teasing, warmed by our own mass action. But, Liverpool are not meant to lose.

If we want any more beer, we had better shut up. During the penalty shoot-out there is a respectful silence – we have probably already gone too far. The instant the winning Man City goal hits the back of the net, the sound is switched off and the programme turned to *Countryfile*, and someone says how Adam is a 'really good farmer'. So everyone, relieved to change the subject, can discuss why he never has mud on his boots and where the puddles might have gone and why doesn't he get a move on and swing that lamb.

Spring creeps slowly up our valley and in the last week we have had 70mm of rainfall, which is not at all unusual. But yesterday the sun came out. The new vibrant verges have splashes of bluebells, and the pond surface trembles with life as the tadpoles jostle for a place in the sun and a tiny frog has his first taste of fresh air, albeit with the assistance of the author.

I put him back where I found him.

But best of all, yesterday our swallows returned, flapping at our bedroom window as the first rays of the sun struck the front of the house. They used to nest in this barn until we made it our home, and ten years later still try to return to the beam above our bed. We close the window and reluctantly they swoop off to renovate last year's nests in the woodshed or perhaps accept our offer of a beam in the new barn.

The house martins that constructed their nest under the north-facing eaves last year for the first time, are back in force, at least two pairs. Last year's nest fell down in the winter, but it looks as if they are preparing to rebuild.

The house sparrows are back in the hole behind the down-spout that we left for them when we re-pointed (not because it was difficult to get at); but, needless to say, all is quiet on the bird-box front.

The moles have been busy re-boring their runs. Alan, not convinced by my argument that their efforts improve the drainage of our fields and that they should be left to get on with the job, stomps off to knock down molehills. He will not trap them though. By teatime the molehills are no more than a memory, smears on the pristine sward.

By breakfast the next morning, and with monumental earth-moving ability, they have rebuilt three or four in each field, shifting hundreds of times their own weight in wet earth. My admiration for the little velvet-suited engineers is not well received by my spouse!

He greets me by the field gate by the barn with a husky whisper.

'What's wrong with your voice?' I ask

'Don't know – haven't got a sore throat.'

Dread seizes me, I am standing in exactly the spot where I had my pressure cooker moment with Gareth, six months before. Since Alan had the all-clear from his chest X-ray, all has been well. Now it is as if all that dispersed, depressurised anxiety that escaped at that moment has gathered itself like a flock of murmurating starlings and is swirling around my head. I know exactly what is wrong with his voice.

He has developed a laryngeal nerve palsy – a well-known

way for lung cancers to make themselves known. It is a late symptom, due to cancer having already moved into the lymph glands around the heart where the disruption picks off the nerve to the larynx, producing a characteristic hoarse whisper. It is a very bad sign – our days are numbered …

But then I suppose they always are – I pull myself together. We have a cup of tea. I ring the doctor from the office so that Alan cannot hear the crack in my voice. The doctor understands immediately and will see Alan after a new chest X-ray which he has arranged for this afternoon.

I drop Alan at dominoes after his X-ray. He is still quite excited – he'd made the pretty young radiographers laugh. But he finds that he can't assert himself in his normal loud, laddie way. Losing one's voice, in itself – I mean really *losing* it, no vocalisation possible at all – is a much greater disability than I ever realised.

I spend the evening refreshing my knowledge about the staging of cancers and death rates.

Surprisingly, whatever is happening in your personal life, the world goes on turning. The clouds roll over the hills and sometimes they part, and the sun comes out.

After a long winter it is all happening again: everything is moving. Even the lazy oak trees are greening. The cuckoo is calling, the cock pheasant is strutting and glinting in the sunlight with a double-*squawk* and a percussive thrill of wings. The woodpecker answers with his own drumming from up on the hill. There are bumble bees overhead and the Orange Tip butterfly flutters over the carpet of white flowering shamrocks on the shallow water of the unfinished pond.

There is a scuffle of illicit nesting beneath the soffits of our roof. Two squirrels, normally too busy to play, are cavorting amorously in the lane, as all around them life springs anew.

Alan's X-ray this time shows a stage IV lung cancer with great big glands at the base of his neck and around his heart. He jokes with the specialist. It is inoperable and palliative chemotherapy is offered and graciously accepted. The meaning of *palliative* is carefully explained to him – 'it will not cure you, but it may buy you a little more time.' Alan's eyes glaze over. When asked if he would like to know the prognosis, he says he thinks not. While he is dressing, the consultant throws me a questioning glance and I look back my answer with a slow, knowing shake of the head.

I have been rehearsing this moment for more than 20 years, ever since we sat in a wildflower meadow in Northamptonshire and Alan asked me, not for the first time, to marry him. I weighed up our projected life together against this moment. It was already too late.

His father had died at 62 of a heart attack, and Alan was a hopelessly addicted smoker since long before his days underground at Bank Hall Colliery, Burnley, where he learned first-hand the importance of mine safety in an environment where the methane levels were never below the level at which the mine should have been evacuated. But he was never blown up – he didn't stop smoking, even then; he learned to chew tobacco!

He was involved in at least one roof-fall and was nearly run over by a runaway train of loaded coal tubs – only his quick wits and agility saved his life. It was in those life-and-death situations that he honed his love of his fellow men and his plain speaking and unwillingness to 'suffer fools', who might get him killed one day.

But it would not be runaway tubs that would kill him.

The bonk

When my first husband and I parted suddenly, 23 years ago, after the previous 23 years of what I thought had been a happy marriage, I experienced huge anxiety – a great explosion of adrenaline. My pulse was rapid, and the weight just fell from my body – 30 pounds in as many days and I wasn't fat to begin with. This physiological overreaction scared me. I thought I might drop dead from the associated electrolyte imbalance. So, I ate bananas, bunches of them, which are rich in potassium. I tried to run off all the unwanted adrenaline, which made me feel less stressed, but then I had to eat more bananas!

My pulse this morning as I sit over my coffee is 120 per minute. I think I need to burn off some adrenaline, and I don't really want to overhear Alan's telephone calls to his grown-up children – I think I'll have a bike ride. My bike is electric but still hard work on our monumental hills. I can't face any breakfast.

I am planning an ambitious circuit as I pump up the tyres and off I set, up the steep hills towards Stay-a-little. I will pass Jac-a-mwn, then turn right just before the village, coast along the ridge, then down the precipitate track past Tynwtra to the medieval bridge at Nant-y-sgyliog. Then a gentle, almost-flat meander along the river to Y Felin where my friend Sadie lives – she came here in the hippy era of the sixties and stayed. That is the plan: about seven miles, but hilly.

'Are you alright?' says a distant voice. I squint – it is Sadie, looking at me the way you look at a sick child. I am pushing my bike past her house.

'Yes, I've got to get home.'

'Come in, have a sit down, have a lie down, you don't look right at all.'

I plod on determined, 'I've got ...' I say something in Latin; I say it very confidently and stomp off in a homeward direction. 'I must get back to Alan.' I make it home, drop my bike in the yard. Alan gives me the same look; he is eating toast. I take it from him and eat it voraciously. He makes me some more and some more. Then he makes me a cup of coffee.

I am feeling fine now, but I can't remember anything – complete amnesia – from riding up our lane until my meeting with Sadie. Alan examines my helmet: no scratches, no mud. He turns me round and examines all elevations: nothing appears to be damaged. He looks at my shoes: no scuffs. And he goes out and checks the bike: still shiny, no mud, no scratches.

I telephone Sadie, 'Do you know what happened to me?'

'You told me – you seemed pretty sure.'

'What did I say?'

'Don't know, didn't understand what you said, can't remember, sounded very clever and medical. Sorry I can't help.'

My doctor daughter tells me it was Transient Global Amnesia, which certainly sounds clever and medical, but I am unfamiliar with the term: it is when a (usually old) person fails to record memory for a time. Was I so preoccupied that I completely overlooked such a strenuous ride.

'Oh no!' says my daughter's partner, who is a cyclist, 'she just had the bonk!' Evidently, this is hypoglycaemia, low blood sugar – what cyclists get when they overdo it on an empty stomach. An example of how something which is commonplace to one group can be a mystery to another. So much for joined-up thinking.

And the treatment? Bananas!

It isn't the more-than-weekly visits to Shrewsbury Hospital that do it. Alan likes to drive there. It takes about an hour, and I drive back. It is not that he is ill after his treatments, but he is tired – showing that one is made of the right stuff, taking every assault and humiliation with devastating good humour, flirting and joking and remembering everyone's name and their personal details takes it out of one ... takes it out of him.

What has made me realise that we have to get rid of our sheep is *separation anxiety*.

If he is going to be ill ... *when* he is ill, I do not want to be 60 miles away, lambing or chasing an escaped ewe, maybe with snow on the ground or floods around Welshpool. I need to be able to drop everything and go with him.

I have explained all this to Gareth, and now he is outside having a good look at the sheep. He comes in and makes me a generous offer. A good price and a kind offer to include the old ewes which he will sort out. He has endless contacts and will probably, he assures me, make a profit selling them to another barmy Englishwoman. We shake hands. Aby is not included.

The grass however does not stop growing – Iolo has the answer. His daughter Bethan has a few sheep, as does his son. They are destined to take over his farm but there isn't really enough land for this to be viable. They will have to rent some more.

I speak to the estate agent and print off the forms, and Bethan shakes my hand. Alan has kept out of these negotiations – he knows it is right but does not want to be blamed.

The Chicken Whisperer is happy to try his hand at Khaki Campbell ducks, and neighbours who keep chickens are only too pleased to expand their flock.

Suddenly we are freed from some of the imperatives of the

farming year, but something strange has happened. For the last few years, I have been writing about our adventures in a local magazine, and now someone wants to publish my articles as a book. This seems to be what has happened in our life – it evolves around us almost passively. Alan appears to be remarkably well. He sleeps more, and I sit with the laptop on my knee and edit. My friend Wendy is illustrating the book, and we spend happy hours at the kitchen table deciding on just the right images, and eating the cake that she brings.

I started writing after I retired because I'd worked out that I would need a back-up plan for when I was too old to pretend to be a farmer, and here I am with the first edition in my hand.

Everything is shifting – we are on moving sand.

My daughter looks at me today and asks me how old I am. I tell her and she says she still can't believe it. That is nice, but I am not taken in. I know that as she looks at me through her eyes, connected to her brain that loves me. She recognises me as the same old Mum – actually the same young Mum. To all the other people of her age, in their prime, I am now almost invisible. Unless, that is, I engage them with my personality. Sometimes, if I am in a hurry, I can flit around town in my cloak of invisibility.

Alan engages everyone with his personality, which is charming, but it is harder and harder work for him.

Aby is also ageing, now the grand old matriarch of our ovine community. In her last pregnancy she was so inactive that we feared for her health – we decided not to put her to the ram anymore. She is retired and, like an Edwardian lady, she has a companion.

Today the air is oozing. It is absolutely still. Not a single raindrop, nor a whisper of wind, but everything is wet, reflecting the mist. Every blade of grass carries pearls of moisture that swell and drop into the soggy ground.

Water condenses onto every surface – all-day dew ... *Dew* (the Welsh for God) that seeps through the sloping fields, that runs down ruts and overflows the puddles into ditches and culverts where it gurgles and giggles to the jingling stream.

The land sings with water. Not falling rain, not today, but water that is a sacrament, a mystery and a power.

Aby is getting to know her new colleague – not quite a friend yet.

The other sheep, including her previous companion, Twts (which is a designation not a name – you know the rule), have all gone to meet the ram. Aby has a new Twts to keep her company. Sheep are not usually happy to be alone, although this particular hand-reared one might well prefer to be back in the kitchen with the dog and me.

That's why today she looks so grumpy. To top it all, the new hogget (who is very undersized) is getting extra rations, which is very irritating to Aby who is on a diet. The new Twts, who is from a neighbour's farm, was an orphan like Aby, so is very bold with humans but still not at ease with Pedro, the dog. She stamps her feet in an unfriendly way when he comes near – he looks affronted (it's early days).

Good times – bad times

Last night was very quiet. I went to listen for owls and night-jars at 4am, but all I could hear was the occasional high pitched *bip* of a bat passing overhead, looking for the last of the midges.

I leaned against the field gate and listened very hard. Faintly there was the white noise of the stream, 50 yards below, a billion splashes and glugs of millions of different, asynchronous frequencies vibrating the air. But above that there was another sound. Above, because it seemed to come from above, but below in pitch, a celestial hum. There was no wind, no traffic for 50 miles, not a plane in the sky. Only drifting cloud over a hazy moon and this strange brown noise (or maybe it was purple). Infinite sound from an infinite number of sources – jet planes over Cardiff, a generator in Machynlleth, the creaking of the trees, dogs in far-off farms barking at the moon (too far away to distinguish individually and too many). Thunder on the coast and the sea lapping on the shore, back doors opening (to let out cats), snoring from upstairs windows and sheep (millions of them) eructating – burping and coughing in the moon shadows.

All these sounds bounce over the Earth, off the sides of houses, resonating in tin sheds and ricocheting off cliffs and bouncing off the underside of the clouds. They can be muffled by the mist and absorbed by the moss and the snow, but they all combine to make the hum of our planet.

We value the darkness of our nights (the lack of light pollution) that allows us to see the brightness of the firmament. Last night I appreciated the stillness of the night that allowed me to hear beyond the silence.

~

Alan is up to something – I know he is thinking about vehicles. We have a relatively new truck which goes very well but is difficult to park. We had it from new, ten years ago. At the time, Alan had just slipped his disc. So when our friend, Rowley, presented us with a gallon of left-over under-seal and the advice that it would prolong the active life of our new vehicle by an enormous amount, you can guess who was sent underneath to do the job. I emerged, spattered from head to foot, with the bituminous emulsion, but since then I have felt a special bond with the five-seater Ranger.

I am also very fond of our camper van, whose value is inestimable, especially as my son-in-law keeps saying he could sell it three times over to surfers in Cornwall for more than we paid for it, aeons ago. Unfortunately, this is always a winning argument with Alan.

What I do not say to Alan is that I like the security of having a vehicle that, if necessary, I can sleep in – I could sleep in the hospital car park. I think quite a lot about the future, and travelling long distances alone is one of the things that worries me most. If called upon to make the trip to Cornwall, to look after a grandchild, and, beset with the urge to nod off halfway up the *Bannau Brycheiniog* in a thunderstorm, it is reassuring to think I could pull over and have a little snooze in a lay-by.

Possibly Alan is also worrying about me, asleep in the back of a clapped-out van, broken down on the M5 near Glastonbury, or brewing up on the hard shoulder, with lorries roaring past, waiting for the AA.

After years of faithful, agricultural service, the White Van goes west, in the custody of the next generation, to retrain as a leisure vehicle – like a middle-aged man having a mid-life crisis. It is painted with psychedelic graffiti, has a roof-rack fitted for surfboards, and swaps the aromas of engine oil, wet dog and rolling tobacco for something more exotic. It's rather nice that it is still *living on the edge,* getting apprehensive glances from respectable people.

I am getting even more suspicious: while I do the shopping, my husband is spending more and more time drinking tea in the office of a local car-dealer. Eventually I am admitted into the conspiracy. They have found me a low-mileage, midnight blue, manual (to keep my mind engaged), low power, economic-to-run, hatchback. I am not impressed by this attack of *little-woman syndrome.* If we ... If *I* must have something smaller, shouldn't I have a vintage Land Rover like Vera, or a gator like Iolo?

'Ah! But ...' says the well-briefed salesman, 'it does have leather upholstery, a leather steering-wheel and heated seats.' We all have our little weaknesses.

I sit in the leather seats. 'They don't burn your bottom on hot days when you are wearing shorts,' I tell the salesman. He nods, 'and they don't hold onto animal hairs – I always arrive at funerals covered in hairs.' He nods again. This man is a consummate professional; he keeps silent as I look around. 'It hasn't got a sun-roof. That's good, I don't like sunroofs. They stop you taking grandchildren into safari parks.'

We leave with our nearly-new, *top-of-the-range* Astra. I have been manipulated and out-manoeuvred. I have just bought a sensible, reliable, small car suitable for an elderly widow – it hasn't even got a tow hitch, but then I can't back a trailer.

~

Alan is telling his oncologist about his only daughter's impending marriage in Cornwall – the specialist looks doubtful. I say that I will be driving and that we will break our journey in the

Forest of Dean and have lunch with my ex-sister-in-law. Alan tells the specialist that he is going to give his daughter away, come what may. I mention that during his last chemotherapy, the thing that made all the difference was the three days of high-dose steroids that gave him so much energy that he was almost back to normal and mowed half the thistles. The medical man is old enough, empathic enough and wise enough to know that there is no point in extending life if the patient cannot do what he needs to do. He writes Alan a prescription for three days of high-dose steroids, he says, to cope with the stress of being driven 500 miles by his wife!

The weather is sunny, with just enough breeze to keep us comfortable but not enough to blow away the marquee, whose skirts wave majestically to the rhythm of Alison's future father-in-law's jazz band. The bride is radiant. The ceremony is good-humoured and light-hearted, conducted by a glamorous retired actress with just the right balance of joy and gravitas to control our horde, who have not seen their father for a while – there is a slight emotional tension in the air. An unusual number of tears are silently shed as he leads Alison down the aisle. I have slight misgivings about the dose of steroids, which is high enough to induce mania in a normal person. But it turns out to be just the right dose for a father-of-the-bride. He does not take over, interrupt the ceremony or try to kiss the lady vicar.

After a magnificent meal, eating more than he has for weeks, he relieves the groom of an unwelcome duty and whisks his daughter around the dance floor in the first waltz – he taught her to dance when she was in her teens, and they twirl as if they have practiced for days. They don't fall over or injure any of the grandchildren break-dancing around them.

Then, like Cinderella, we depart.

We are staying in a holiday home for a week.

When I am home in Wales, I spend many a happy hour musing on possible sites to land a helicopter – specifically, an air ambulance. Basically, there is nowhere within two miles that is flat or not festooned with power lines or patrolled by ferocious cattle. The only flat place on the farm is where Alan has built an enormous bonfire which was too dangerous to light all last summer and is now too wet.

We are in Lostwithiel in Cornwall, on the Fowey River. There is a colossal roar – so tantalizingly loud that Pedro and I must go and investigate.

Fifty yards from the house, we find the source of the row: a Cornwall Air Ambulance, come to pick up someone from the adjacent medical centre.

We watch, from a distance, as they are being loaded.

It's already dusk but how wonderfully flat it is, and not an overhead wire or a bull in sight.

It starts to roar again and the rotor blades, which were drooping, start to rotate faster and faster and get flatter and flatter and louder and louder until one thinks it can't try any harder. But it does and, as Pedro sinks to the ground and covers his ears, it lifts lightly up and turns to face us.

It flies directly at us and seems to smile (must have seen the camera) then rises up in an aerial pirouette, and sets off towards Plymouth. Good Luck!

Pedro seems off-colour. He's been indoors quite a lot in the last couple of days. I'll give him a bit more of a walk before we go back to cook the supper. I pass another dog walker. We chat, and I ask her if there is a vet in the vicinity – there is a good one just up the road. We'll walk up there and find out the name, phone number and opening hours. We walk on but there is no sign of a surgery. I ask someone else – it's about a mile, up the hill.

Pedro always wants to be ahead. Suddenly today he is lagging behind, and I realise he is not going to make it. We turn back. He is panting now … nearly back at the holiday home. It has got dark but there are street-lamps. We cross the road by the turning to the house. Suddenly Pedro slumps down onto the road – right in the middle. I cannot move him. There is no traffic at present. I give him a few moments then try again. There is no-one about. I try to lift him, but he is far too heavy, so I pull him to the side of the road, apologising frantically to him as we go. Then I leave him draped in my bright-coloured anorak and run as fast as I can back to the house.

Alan brings the car round to where Pedro lies, his chin resting on the kerb, but even with Alan helping we can't lift his dead weight into the boot. I run back and get his large, plastic dog basket and we manage to slide him into it, then we can get a good grip and with a heave lift him into the low boot of the hatchback (thank God for the steroids!).

Back at the house, he is lifting his head and seems to know us – he has some water.

We find the vet's number in the *Useful Information for Guests,* and they answer promptly – they are just closing but if we come immediately, we will catch the vet.

We have to carry him in on his dog basket stretcher. He lifts his head and looks sadly at the vet who pulls down his lower eyelid to reveal just how severe his anaemia is. His pulse is racing. He has had a massive internal haemorrhage and there are signs that this is all due to a tumour. How did we not know this was going on? There really is no hope. It is settled there and then. I hold his paw as it is shaved and injected, and he slips away – our best dog ever.

It's been a difficult winter, what with one thing and another. Several of our good friends have crested that distribution curve for life expectancy and done what we will all do eventually. Because of this it hasn't seemed right to talk about the passing of our dog.

However, Pedro was such a special individual he deserves a moment's thought.

He had such a subtle way of attracting the attention of everyone he met, that rescue cur, that eloquently and silently persuasive dog. He hardly ever barked – he didn't need to. The only time he ever howled (except when the grandchildren taught him to one Halloween) was on Remembrance Day. On the television, a lone bugler played *The Last Post*. It is always an emotional moment and quite spontaneously, one year and every year since, at the very end, Pedro lets out the saddest wail. One howl, completely in tune. It may be a trick of the pitch but it touched the core of everyone who heard it.

He understood our everyday chatter.

'Not in front of the dog,' Alan would say as we discussed the *possibilities* of an outing. 'We don't want to disappoint him.' In reality, it had probably been his idea in the first place. He had a way of inserting thoughts into our heads.

He would fix me with his stare and then glance at the object of his desire. (Whole systems of possibly dodgy psychoanalysis are based on this method of communication. It has a jargon name which escapes me, and is undoubtedly out of date anyway.) Not all humans are wired up for this system. But I am – I am reactive to everything I see. People like me flit around if they are not careful. They are so distractable that they pick up on all sorts of facts and stimuli that more focussed individuals miss. His glance would lead me to the path into the woods; his ball on the shelf makes me think 'let's play out'; my crook, and I remember that I wanted to look at the lambs; wellies lead us to the stream or the beach (he loved the beach). He knew exactly how to introduce his thought into my head: feed the sheep, collect the eggs, walk the dog and don't forget we are taking next-door's dog today. It's six o'clock *(I know it is, I'll feed you in a minute!).*

Come on, time to set off! He is sitting in the driver's seat with his paws on the steering wheel. If he had a watch, he would be tapping it.

When Pedro first came to us, Alan insisted on feeding him – 'dogs have to know who's boss, whose dog they are, who's in charge'. For all that, it soon became apparent that Pedro was *my* dog. I was his human. We were on the same wavelength. Not because I am bossy (although I am) but just because we understood each other.

The thing about dogs is that they also communicate on an emotional level – with irresistible sadness when they don't get their own way, and uncontainable joy when they do, and joy is catching. A walk in the woods or a romp in the snow with a happy dog can elevate the meanest mood!

Pedro was not so good with cats, but he would have loved a kitten. Cats assert themselves by emotional blackmail. Pedro thought, *they are not team players.*

He was a proper family dog. Good with children, sheep, and lambs. He was fascinated by baby creatures. He'd bring them in but never hurt them. Dogs like him were crucial to our distant ancestors, I bet, in the domestication of other species ...

In his previous career he had been a professional lap-dog. But, like the ballet dancer that got too tall, he just got too big – no lap could accommodate him.

He was an athlete. He jumped over fences and gates without touching the top. At least he did after he had unzipped himself on barbed wire so that his skin opened like an unbuttoned beige cardigan, flapping as he wagged his tail. He was oblivious to the injury – a stoic (adrenalin again). He was so excited, I could not have sewn him up, even though I had the kit. He just ran round me in circles. Off we went to his new friend – the vet.

Where will we find another such astute farm manager or a dearer and more faithful friend?

Freezing February

Last autumn it was too dry to burn the brushwood from our extensive hedging operations.

Now, when most years we have snow, I've been farming in my shorts. After the frosty start, we've had the hottest February days since records began (here anyway). There are wildfires on Saddleworth Moor, but here the ground is still a bit soggy. So Alan announces that the conditions are right for a bonfire! We have a pile of brushwood 20 feet across and 10 feet high. It is probably packed full of dozing hedgehogs – there are heated discussions.

As we had a spot of bother with our last big fire, our friend Gareth takes it upon himself to supervise us, bringing the grab on his big tractor (always exciting for us). Ever mindful of the wildlife, the grab will enable us to deconstruct the heap little by little – or a lot by a lot – and add it to the fire in smaller amounts, out of which the hedgehogs will tumble and roll away to safety.

There was a shower overnight, so it is slow to start.

However, after a little chemical persuasion, we have a spark to work with. It flares 15 feet into the sky with a roar, and Gareth drops an industrial load of brushwood onto it from a great height. Smoke billows into the blue sky, at first black then a copious, swirling column of white steam and smoke, reaching up, vertically into the stratosphere – visible from the moon.

Now we have a decent bonfire, which Gareth and Alan can watch, leaning on the cratch. Satisfying to watch, but not we hope from above (none of us can quite recall the exact regulations).

It burns all night, and no hedgehogs were injured in the making of this fire!

By next day it is manageable by a retired lady with a pitch-fork, and we are ready for the spring. And (you guessed it) it's raining!

Now I have a heliport once I've moved the cratch, and I'm wondering about how to make a giant 'H' on the ground.

The other day I watched a quad-bike on the far side of the valley, feeding a large flock of sheep by driving in a large circle with sheep nuts flowing from a sack on the back of his bike. The sheep chased him and formed themselves into a perfect O! It was there on the hillside long after he had gone (writing with sheep – there's a thought).

Alan is having a fearfully expensive, experimental, new treatment with an immunotherapy drug. The results, we were told, are very promising. When he starts it, I ask the consultant how promising? 'So far,' he says, 'there is a definite increase in life expectancy.' Alan goes back into the examination room to find his shoes. 'How great an increase?' I ask.

'Oh,' he says absently, 'several weeks'.

Today Alan spoke to his elder son on the phone. I thought he sounded slightly grumpy – not like him.

Five hours later his son rattles the French windows. He has driven up from Cornwall, so worried is he about his Dad.

Famously youthful and stubborn, Alan has refused for nearly two years to make any concessions to his inoperable lung cancer, and everyone is beginning to think that he might just beat the predictions.

This week he has changed all the blown bulbs in the house, checked his will, caught the mouse that stole his chocolate, MOT'd the truck and got the puncture on the car fixed, and entered his motorbikes into an auction.

When his son arrives, Alan insists on searching the local off-licences, buying five bottles of spirits (unusually expensive and sophisticated – all the special tastes he remembers from his travels in Europe. He can't manage a pint any more). He lines them up on the kitchen table, opens one and tastes a little Calvados, mends the oven door, has a smoke and goes up to bed, sits down on the side of the bed, and dies.

What a way to go – Alan did everything on his own terms.

And now ...

I am alone. I am gazing out of my office window.

I spend a lot of time looking vacantly *over* my computer and out of the window. Within six feet of me and at eye level, there is a very fine mistle thrush – my new friend. I don't know if he can see me, but if he can, he certainly isn't bothered. I suspect he is more interested in the handsome reflection that is eyeing him quizzically.

I've put out a fat-ball for him. I don't buy meal worms since a young visitor was caught putting them on her muesli.

It is his cousin, the song thrush, that is the first to sing every morning, piercing my slumber just before dawn. He sits at the very top of the tallest tree.

Now what's this? On the bank, stealing from my new friend's fat-ball. It is a bank vole. Gosh! He moves fast, flitting all about the bank. Soon the plants will have grown so much that he will be sheltered from the crucial eyes above, the ones that threaten him: those of buzzards, barn owls and tawny owls. But watch out! There's a stoat that visits this bank, and next door's cat. Everything is getting much braver since our last cat was repossessed by my eldest daughter, once she finally settled down. And, of course, since our dog died. I forget sometimes, when there is a noise in the night – *just Pedro*, I think.

I am trying to write Alan's eulogy – the vicar will read it. The family have gone away and will be back for the funeral, and the phone never stops, which is very useful. Old friends remind me of funny stories about him, that I had forgotten or never knew.

A friend sends me a photograph of him in his gabardine macintosh, goggles and black beret, speeding along a mountain chine on an old-fashioned motorbike, dressed like someone from the wartime French resistance. I am told of sporting heroism: 'Did you know he once played in goal for my hockey team? A forward scooped up the ball with his stick and fired it high towards the net, which might have been a goal. Alan couldn't get his stick or his foot to it, so he headed it! You must include that,' insists the elderly man on the phone, 'it was Alan all over.'

Goodness! It's getting dark. The bell outside has just rung – the large ship's bell by the front door that people ring to get our attention. It is Bethan. She is standing in a pool of the light, holding a bundle.

'There aren't any cars. Have they all gone?' she asks.

'Only till the weekend. They all had to go back to work. What have you got in your bundle?'

'It's an orphan lamb – Dad thought you might like to adopt it.' I lean in and pull down the old towel that covers it. 'We were going to put it in with the others – we've got several on the feeder in the barn but it's a good little female ...', she adds. 'You could keep her.'

'How beautiful ... Bethan, that is so kind ... so thoughtful ... so tempting ...'. I am thinking, that is just what I *want*; everything will start all over again. Next I'll get a puppy. But is it what I *need*? If I accept this wonderful gesture, life will go on just the same, only I'll be struggling without Alan. He will always be missing.

Perhaps what I *need* is something different …

~

Widowhood was sudden, though always half-expected because no-one can expect to be happy forever and I did know he was ill, although he pretended not to be. Now my children, who are suddenly definitely grown-up, worry (and probably moan) about me at least as much as I do about them. I keep wondering why people are being so nice to me, then I remember.

Suddenly I can do whatever I want, although I don't really want, but I do it anyway, because I know I must. I have been very inactive since Alan's illness, but yesterday I climbed a mountain with a group of strangers to look at historical sites – one of those Welsh mountains that is really only a huge hill. I was interested in the archaeology. The others seem to be serious, serial walkers – there was talk of Kilimanjaro! It was cold and steep, and they walked very fast. I got short of breath (which never used to happen and probably was not the altitude), and I hobbled a good deal on the way down. The leaders took it in turns to hold back and walk with me, looking worried, but I walked eight miles and I didn't die.

'She's got rid of those breeding ewes – we won't have to look at any more pictures of slimy new lambs!' I hear my friends say.

Not so. As with all things in the Garden of Eden, Mother Nature will have her way! Bethan now uses our land every year, and although I still have Aby, I resist the temptation to adopt another. At present, Bethan is grazing her adolescent ewes but is learning just how fragile is ovine virginity!

The first two happy accidents are discovered on Wednesday – two fluffy, lively, white lambs, standing to attention next to their pleasantly-surprised mothers.

Bethan leaves them in peace in our top field and brings the others down to the fields around the house. I am out. The first thing I know about what is going on is when I am eating my lunch in the sunshine and hear a strange *baa* – like a child imitating a sheep. I go to investigate, and find a bewildered young ewe with tummy ache. But something is wrong (didn't I say that *anything that could go wrong, would go wrong*?). No, that was my Dad, and he called it something else! It's the first rule of rearing anything. She is agitated (as well she might be), and not progressing in her labour – I work that out at least. I try to catch her, which only reminds me of another reason why we decided to stop lambing in the first place. I phone the shepherd, who phones Iolo, her Dad, who borrows her dog, who comes and catches the ewe. Remember, these are classically-reared, proper farm ewes who are not trained to follow the magic pot, nor can they be marshalled by a trusted, four-legged assistant.

Bethan's dog drives the ewe into the pen, and Iolo and I extract a very shocked, large and strangely khaki-coloured lamb with a swollen head, from which protrudes a large, congested, blue tongue. She is initially disinclined to breathe. But, after all the swinging, poking, massaging, passionate entreaties and expletives, she at last relents and takes a mighty gasp. We sigh with relief. Mother Nature, whose baby this is, is not about to be outdone at this stage!

Iolo is muttering that he doesn't know how it could have happened. I tell him he should never underestimate the cunning of a lovelorn female. The next arrives a few hours later and is still a bit wobbly; the mother is much calmer.

After her day-job, the shepherd arrives to check the rest: two more wayward adolescents are identified, to be collected tomorrow and taken to the main farm. But, guess what? What's this? A *fifth* healthy, vigorous lamb.

Now, it's getting dark and I'm going out to look for number six!

I told you at the beginning that this is a sheep farm and that it has its own agenda: I know it is trying to draw me in again.

What a treat it is for me to have some field lambs to fuss

over. But what strikes me most is how big and healthy these lambs are without all the extra food and care that would normally have been lavished upon them.

I know what Gareth would say: 'Young stock!'

~

A cade lamb, orphaned at birth and brought up by a silly woman and a clever dog will not be like other sheep. Aby has had her portrait painted, has been photographed for magazines (not *Hello!* but she was the centrefold for *Border Life*). Basically, she is a celebrity – the ovine equivalent (in my mind anyway) to Joan Collins, and also looking very good for her age.

Yesterday in the shed, I found a length of rope with a metal ring attached, that I thought was an old calf halter. And I used it to tie a great sheath of brushwood to the wheelbarrow so that I could overload it like a Greek donkey and wobble to our 30-metre-long, heavy-duty compost heap – our special habitat in the woodland. This is a haven for dozing hedgehogs, nesting wrens and the innumerable wood moulds and fungi that live in our little piece of temperate rainforest. Aby comes with me for the walk.

On the way back I wonder if Aby is too old to be halter-trained. After all, she does identify as a cross between a dog and a human – a category that EID Cymru, the guardians of sheep identity, refuse to recognise on their annual sheep and goat inventory (give them time). As usual, she is walking to heel. We stop and I pick up the halter from the barrow and thread it into a noose configuration: simple. Aby looks interested, lifting her head up to look at the circle of rope that I hold in front of her. Without thinking I just slip it over her head. She is pleased with this new award and sets off to show it to Tex, her latest companion, with me still attached to the other end. As she ploughs on through the mud, I am left behind. I tug on the rope to demonstrate the principle of the choke-chain. This is a noose, not a halter. Halters fastens around the head, not the neck. The noose tightens, she pulls harder. Not wanting to hang her, I let go. She heads up the hill. Her deadly pendant is dragging in the mud and looking for something to grab hold of, so that it can strangle poor Aby or break her neck.

I give chase. I grab. I miss. I grab again (this is fun). I catch it as it circles her neck. She accelerates and swerves. I lose my footing and describe a wide arc, landing on my back in the mud (I think 'fractured femur, hospital, covid' but I don't let go). I am not in pain. Aby is no longer pulling. She is lying on her back with her feet in the air. What a piece of luck – she is cast: immobilized by that primitive quirk of sheep neurology. I have time to pull myself together and remove her rope noose before turning her 'on' again.

Our relationship may take longer to rectify. She stomps off, feeling humiliated and totally let down, and has been firing withering looks at me ever since.

Like the car that rolls and has a dent on every panel, I am wet and muddied on every surface.

Still ... serves me right.

The future

I am looking at the view of Plynlimon from the top of our hill, and it is a political picture that I see. I have a friend who loves layers in the natural landscape, but here I am looking at four man-made layers. It just feels wrong.

From the top, there is over-grazed mountain (not a tree in sight); then wind-farm on desolate peat-bog; then impenetrable, monoculture pine forest; and, finally, mechanised farming in the valley.

In contrast, the farmland we tend here in our small part of Wales is designated by the Government as 100% habitat (which is probably true of most places if only you know what you are looking at and whose habitat you are talking about). But it's official, half our land is 'oak and wild hyacinth' (bluebell woods to you). Ancient woodland that was felled after the war for pit-props for the economic recovery, and grazed until 2006, when the Government (with unusual wisdom) offered us a modest grant to replant – and, more importantly, to exclude grazing for 15 years.

It's been more than that now (time on the land goes much more rapidly than bureaucrats imagine). So, for nearly two decades, ten acres of this land, nestling under the old hillfort, has been spared the ravages of the hardy native sheep that we love, but whose mission is deforestation.

I never understand why folk get so enthusiastic about protecting the bleak moorlands of this area that are scoured bare by unnatural numbers of hungry sheep when, if left to its own devices, this land would be broad-leaf woodland bursting with wild flowers, insects, songbirds and little furry creatures.

So here we are. Our saplings – oak, hazel, rowan, aspen, alder, wild cherry and holly – wrestle with self-sown birch and willow and the creeping shoots of blackthorn and hawthorn which insinuate themselves from the old hedgerows. They were planted naturalistically (not in rows) – not to confuse the tree-counters from the Ministry (that was inadvertent, another happy accident) but to give them a head-start and to make the wildlife feel at home. In the wet gulleys the alders are already seven metres high in places, and alive with siskins in the winter, flocks of busy little green birds. I don't like to embarrass them, the trees, but they are sexually mature with lots of little cones, and the rowans have berries and there are wild cherries. They are visited by bullfinches, flocks of goldfinches and long-tailed tits. In warm summers we can now harvest sacks of hazelnuts, and lose the children for hours picking wild bilberries from the forest floor. We call them whinberries.

Some of the new oaks are twice my height (I sound like a proud parent), and in the spring and early summer the foliage on the new growth is bright red.

Our newly-invigorated ancient woodland is very young, and we will need to maintain the glades and open areas. It would be nice to reintroduce a charcoal burner or an oak tanner (now sadly extinct), to maintain the woodland clearings where the meadowsweet can grow – as it does now in the floor of our little dingle.

Imagine the vanilla perfume, the hum of bees, the tintinnabulation of the stream almost hidden by the over-reaching ferns.

Between the trees, the moss still grows, once harvested for florists. I sink into it. It is like green snow.

I am still exploring this new landscape, which is constantly changing. Already it looks wild and natural, but cut back the undergrowth a little and you will find signs of quite sophisticated engineering from long ago, built by hand from river stone using only a shovel and a mattock, like the one we found with the metal detector.

But beware invaders when you clear ground. Where Alan and I dug out a hidden culvert in the spring to unblock it and release the pond that had squatted along our track, we now have a bank of rosebay willowherb.

I still can't get used to being singular.

What amazes me is the variety of plants and animals that show themselves as the year progresses. Every week the micro-landscape changes as the colours and shapes reflect the constantly changing balance within the ecosystem. As taller plants – like the ferns, the myriad tall grasses, the foxgloves, meadowsweet, and the parsleys – grow up and take the light, the undergrowth of smaller plants – the mosses, shamrocks, wood anemones and bluebells – having flowered while they could, are obscured, and I must wade, shoulder-deep in a tangled profusion of humming, scented, sometimes prickling jungle. The lushness and fertility of it all just knocks your socks off as a fleeing woodcock rises, squawking, just feet away.

The other half of our land (plural again, as Aby walks beside me – she has forgiven me) is officially 'severely disadvantaged' and 'unimproved' pasture – what a cheek! We work hard to maintain it without recourse to chemicals or artificial fertilisers. We hack down the bracken and dig out the gorse and cut the thistles just before they seed, but not right to the ground and not around the edges – that's for the caterpillars. We kick down the molehills and mend the fences, and Bethan's sheep and Aby do the rest once the orchids have seeded.

The trouble with Nature is that it runs away with you!

You give her (Nature that is) a small area of hillside to play with, and before you know it, she has barricaded herself in with thorny thickets and hidden natural earthworks, molehills under layers of slippery bracken and ankle-breaking, knee-jarring pitfalls made by rabbits and badgers (did I tell you I've just had a new knee?). The whole area is now dense undergrowth, criss-crossed by looping, flailing brambles and willow whips.

Ah, you say, *that's nice – good for biodiversity!*

But is it? It's not as simple as that.

So concerned are we that we have called in an expert, a conservation hero – Bionomic Man, Rob.

He'll know what to do.

And he does. We slither and scramble up and down our precipitous banks as he introduces me to species of which I was oblivious. Pleased to meet you! He demonstrates how to distinguish between our six common ferns, and how to make sense of some of the mosses (not easy as they mainly have Latin names, now recorded in my roughly dried notebook).

The grasshoppers that jumped out and away wherever you trod in our meadow last summer were green and there were lots of them. That might make you think that they were Common Green Field Grasshoppers. But, with talk of global warming, climate change and species in all the wrong places, I have been inspired to have another look at my photos and to try to be more precise in my identification.

It seems my first impressions may well have been correct, and this confirms me in the belief that things should be named for what they are. Although in this case I had such difficulty in photographing them clearly that Brown Kneed Elusive might be a better name.

It will be fun giving the mosses proper English names – greater and lesser shaggy Christmas tree moss for a start. The diverse populations of liverworts, lichens and fungi will keep me occupied for the rest of my life.

More importantly, Bionomic Man shows us what we might lose. In this emerging woodland, already we are losing the avenues of open ground for butterflies and moths, insects, and bats. The paths that remain are steep-sided – like canyons through the trees, without the gently-sloping edges, clad with brambles and climbing plants like traveller's joy, needed by moths, butterflies and pollinating insects. There is a lot to do.

Without grazing and trampling, overgrowth of bracken is alternately shading and insulating the great anthills on the sunny bank so that our ancient neighbours, the huge colonies of yellow meadow ants, cannot control their temperature as efficiently as they have done for thousands of years.

The carefully-placed owl box overlooks dwindling areas of decent hunting ground for barn owls as the vole habitat is being eroded by blackthorn (probably the reason they haven't used it of late).

Come to think of it, I haven't heard the churring of a nightjar for a couple of years now. They like newly-cleared forest where they nest around the edges and hunt over the new growth, like large, nocturnal swallows or swifts. Our trees must be getting too big for their liking.

Our mature oaks, it turns out, are only about 100 years old, adolescent almost. Not nearly gnarled and hollow enough to provide adequate nesting for all the bats, like the ones currently breeding in the bathroom ceiling, and the pied and spotted flycatchers that come looking for nest sites every spring. The new barn will need a large, well-insulated roof space to deal with the housing shortage! A new owl box is planned for the old barn, and a safety rail for the swallows and martins as the roof purlins are too steep (the nests tend to fall off).

It seems I'll have to start production of more bird nesting boxes – open-fronted ones for flycatchers. Did you know that spotted flycatchers like to nest near buildings. I should think that dates back to the time when farm buildings had adjacent muck-heaps – perhaps (no, that would be a step too far!).

Now I've got to go and plan the new pond. Alan and I never did get round to it – it was always too wet. We've had an unusually dry spell and Meurig is coming with his big digger and new laser level – he'll do a proper job. I know exactly where to put it and its adjacent compost heap and en-suite woodpile. Our lizards need somewhere to bask with easy access to an air-raid shelter. We've got just the rocks we need, that were the base of the old Dutch barn that fell down before we arrived.

Seriously, the ecological survey was very helpful. Now we realise that managing an area for wildlife is not a passive exercise. Those who think that any form of re-wilding threatens the rural lifestyle had better think again. We no longer work, day and night, to squeeze as much product from the land – lamb, pork, timber, moss, potatoes, leeks, oats, eggs and soft fruit – as our predecessors did.

The trees grow, steadily absorbing carbon and holding back the water, and we will try to help if we can, to re-establish this temperate rainforest. Although we might only be cutting some of our hedges every other year (to get more berries) we will be generating lots of other activity. Our oak won't be ready to harvest for several lifetimes, but the woodland is not totally unproductive. Already the trimmings from fallen trees, blocking tracks or from those felled because they threatened buildings, produce more fuel than we can use ourselves. And we have already allowed some limited grazing in the wild area for the sake of the ants and the owls. Now the hazels are producing nuts – and wait until we get our bee hives and truffles going!

What we really need are a couple of old-fashioned, longhorn cows to control the brambles and maintain some sort of access (the RSPB have them on their reserves). As you know, sheep tend to get caught in the brambles. The cows won't even need stock-proof fences. They can have sat-nav collars that give them a little buzz if they stray too far in any one direction. We don't see deer (too much hunting in the past). But maybe we could have two or three weaners in the oak wood – I'm going to read about wild boar. It's all a question of balance and not upsetting the neighbours or spifflicating oneself!

I think of what will happen when I am gone, when the house appoints my successors: how will they live? I hope this will be a paradise of moisture-loving plants and trees. The red kites, buzzards and dastardly goshawk will still wheel overhead, and a pair of ravens beat their way purposefully across the sky, with their characteristic *cronk*.

Over the new pond, the hirundines will swoop – swallows, house martins and swifts – on the clouds of midges, as the dragonflies patrol the banks and damselflies and demoiselles hover in the shade by the outflow. Under the great oak trees the pied flycatchers, small black and white birds, will still perch on dead branches before fluttering up to grab an insect on the wing. A treecreeper, with its downcurved beak, pretending to be a mouse, will scurry up the rough bark of a tree trunk, looking for insects. And a nuthatch will peck at a seed jammed in a crevice, hanging upside down with a *rat-tat-tat*. You will hear the drumming of the woodpeckers, the *squark* of the jay and a 'penny spinning on a marble tabletop' – the sound of a wood warbler. There will be flocks of goldfinch, siskins and long-tailed tits, and all the other tits, eating the profusion of seeds and berries maturing at different times from all the different flowers and bushes. The wrens will forage in the leaf litter, and shout from cover when someone passes. And snipe will be almost invisible amidst the dead bracken along the wet pond edges and in the dingle.

The dell will buzz with bees. Hoverfly and small Pearl-bordered Fritillary butterflies will pause for an instant on the bramble flowers. Diving beetles will harass the tadpoles in the top pond. There will be Marsh Fritillaries on the beautiful buck-bean flowers in the bog by the new pond, where brown trout fry already dart about in gangs. Above, a kingfisher perches in the willow on the little island because the children have all gone now, and their haul of tadpoles, newts and dragonfly nymphs have all been safely returned to the water. In the new pond there will be a kerfuffle as a huge, black cormorant rises like a phoenix from its surface, quite startling the heron.

The new owner will be outraged by both these incursions. The thistles and nettles will still not have been cut, and Small Tortoiseshell butterflies and Red Admirals will mill about the thistles in the remaining rough pasture in the sunshine. In the hedge up the hill there will be Green Hairstreak butterflies on the hawthorn.

In the afternoon on sunny days, the lizards will bask on their little stone mountain – and what's this? – a Wall Brown butterfly. People will sketch the myriad flowers, ferns and beautiful lichens. In the autumn they will come to spot the mushrooms and spectacular toadstools, fruiting bodies of an amazing variety of fungi. They can have foraging workshops (best stick to blackberries, nuts and whinberries).

There will be moles and rabbits, squirrels, stoats and pole cats, hares and maybe a glimpse of a wild boar or a visiting otter (there is room for a beaver in the dingle). It will be a place that people will want to visit and photograph. They might wonder how a Great Dawn Redwood tree managed to grow in the centre of this small area of rare temperate rainforest. *Metasequoia glyptostroboides,* as Alan would have proudly addressed his favourite tree, had they met (it was the only Latin name he knew). If they are very observant, they might even notice the remains of an ancient but once very substantial tree protector tangled around its base. Don't worry, this tree has no mate – it will not take over.

The new custodians can fill the house with paying guests, walkers and cyclists, and sustain themselves by taking no more from the environment than they give back.

You never know.

Watch this space ...